Principles of Spinning

Principles of Spinning

Carding and Draw Frame in Spinning

Ashok R. Khare

CRC Press
Taylor & Francis Group
Boca Raton London New York

CRC Press is an imprint of the
Taylor & Francis Group, an **informa** business

First edition published 2022
by CRC Press
6000 Broken Sound Parkway NW, Suite 300, Boca Raton, FL 33487-2742

and by CRC Press
4 Park Square, Milton Park, Abingdon, Oxon, OX14 4RN

© 2022 Taylor & Francis Group, LLC

CRC Press is an imprint of Taylor & Francis Group, LLC

ISBN: 978-1-138-59658-0 (hbk)
ISBN: 978-1-032-10513-0 (pbk)
ISBN: 978-0-429-48656-2 (ebk)

Typeset in Times
by KnowledgeWorks Global Ltd.

Contents

Preface

Over the years, though the various functions of carding machines have not changed, with the rapid development in technology, the design of a modern card has adopted many new concepts in improveing these functions. Even then, the proverbial phrase, "To card well is to spin well", still holds true. However, a modern card really helps in making the card work well with its new attachments.

In this book, therefore, ample emphasis is given to several of these attachments. For example, the licker-in and licker-in zone are very much modified. Whereas the conventional mote knives have been totally done away with, newer versions, along with the use of additional licker-ins, modified hoods, suction to remove flying dust, etc., have much improved the cleaning efficiency of a card. This has been adequately assisted by the chute feed itself. This is because the licker-in does not have to exert any additional effort to open up the compact lap and then carry out the cleaning process. The loose cotton feed received by the licker-in also enables it to carry the job of opening the tufts further.

This has greatly reduced the burden on the cylinder in carrying out the basic function of individualization. The provision of carding segments and improved flat designs and wires greatly assist the cylinder, therefore, in thoroughly doing the carding of the stock. The cylinder speed itself is yet another area which has considerably reduced the fibre load on its working surface. The cylinder of a modern card runs in the region of 400–450 rpm, and on some of the latest versions it has surpassed these limits.

Roller doffing is not new to the latest generation of cards; the Platt brothers were the pioneers in introducing the Crosrol Verga concept. But the design and the metallurgy of the doffing and redirecting rollers, and the pressure building on crush rolls have gone a long way in improving their functions.

Many systems have been automated. Thus, the card has many sophisticated stop motions and is equipped with many online electronic controls for adjusting the settings to improve its basic functions of removal of the trash and individualization. At many suction points, the arrangements are made to automatically suck the liberated trash and fine dust. Even the bifurcation of dirtier licker-in droppings and comparatively whitish flat strips are collected separately so that their subsequent processing becomes easy. The whole card is fully enclosed so as to avoid any card-room contamination.

Sincere efforts are made in this book to highlight most of the important features of the modern card. Apart from this, as mentioned above, the basic functions of the card remain the same and, the functions of the conventional card are explained in the traditional way.

In regard to the drawing frame, the focus is always on the drafting system, and therefore, here, too, some effort has been made to highlight the features of modern draw frame, especially the drafting system. Basically in drafting, two things are important—roller diameters and their setting provision and roller weighting. While pneumatic loading has substantially improved the roller nip, stronger bottom rollers with helical grooves have also helped in doing a similar job. On all modern draw frames, it is now possible to correctly set the rollers with precision, the credit going to the arrangements provided without involving the traditional method of having to insert the flat gauges. Apart from this, there has been improvement in the types of cots available for use. All of them have improved the coefficient of variation of drawing sliver (U%).

The latest versions of modern draw frames have added regularity controls on the principles of *open loop* and *closed loop*. This has been reported to improve the drawn sliver regularity. It is thus possible to produce a sliver with less than 2.5% Uster value. As usual, the improved stop motions, bigger cans, and higher working speeds all aim at quality and output rate of the machine. Here again the whole drafting area being enclosed under a suction hood keeps the surrounding atmosphere clean.

The useful chapters on defects in carding and draw frame material and related calculations on both of these will almost complete the knowledge of these two processes. I hope that this book, like my earlier ventures, will be well received by the student community.

Dr. Ashok R. Khare

List of Tables

Acknowledgments

Since I was a college student, I dreamt of becoming a teacher. A few of my teachers were my idols then. I feel proud to mention their names. This is because later, when I became a teacher in the same college after several years, I was always inspired by remembering them, their style and philosophy of teaching, their sincerity and devotion to the profession and their skills in making difficult things look simple. All of them had good industrial experience and were therefore able to share their knowledge with their students.

Prof. D.B. Ajgaokar, who later became the first Principal of D.K.T.E. Institute Ichalkaranji, encouraged me to venture into writing books for students. After my first success, he continued his encouragement and support for writing more. This is how I was able to take up this vast task of writing a book series on spinning technology.

The late Prof. M.K. Naboodiri had been my philosopher and guide, and very few know I had a family-like relationship with him. When any problem was posed to him, he would never offer haphazard answers. He would meditate on it and then come up with a logical solution.

Dr. V.S. Jayram was my first teacher when I started learning about textile technology, a field which was unheard of until the completion of my school days. I greatly enjoyed the way he taught spinning. I must admit that the credit of my continuing with textile studies goes solely to him. As a student, I never missed his lectures. I must also confess that his style of teaching inspired me to come back to my college as a teacher. When I requested that he edit this volume, he very gladly accepted and completed the work. But most unfortunately, within a short period thereafter, he died. I will always remain indebted to Dr. Jayram.

Dr. S.G. Vinzanekar, my mentor during my whole career as a teacher, was always a source of inspiration for giving me several opportunities to learn things. When he would entrust me with any job, he always used to fully back and support me, irrespective of whether it was a success or otherwise, and even in odd circumstances. He was my guide for my PhD work, and there, too, he became my senior philosopher friend.

I am extremely thankful to Dr. Patil, Director CIRCOT (Central Institute for Research on Cotton Technology), Mumbai, for helping me from time to time, and for making the CIRCOT research on cotton technology available. Thanks are also very much due to two giant and reputed machinery manufacturers: Trumac-Trutzschler and Rieter for providing me with beautiful diagrams supporting the theory of carding machines. Without their help and permission, this book would not have had real worth, which I feel is what it is now.

The base of this book appears to be similar to the *Manual of Cotton Spinning* book series published long ago by the Shirley Institute, though in treatment of the subject information, it differs. Even so, I profusely thank The Textile (formerly Shirley) Institute, Manchester, London, for giving me the inspiration to write and add some useful information to the ocean of textiles. Equally important was permission from Elsevier (former Butterworth publications) who permitted me to refer to their book *Spun Yarn Technologies* (by Eric Oxtoby). I am greatly indebted to them. I hope, in the present form, it is still useful to students. Thanks are also due to my well-wishers, who directly or indirectly helped me in this venture.

Last but not least, I would be failing in my duties if I did not mention the name of my father, the late Shri Ramchandra Narayan Khare, who since my childhood groomed me to become a good student, and later to be a good teacher and a good citizen. He himself was a born teacher and expert in child psychology. When I was thinking of leaving my job in the industry to begin a teaching career, he asked me only one question: "Would you take up this career with all sincerity and dedication?" He also implanted in my mind that if I am to be a good teacher, I always need to be a good student. I have never forgotten his words during my whole professional career as a teacher.

For my ability to author this book, the credit goes to all these great personalities as, in some way or another, they have been instrumental in making me as I stand today.

Dr. Ashok R. Khare

About the Author

Dr. Ashok Khare has graduate, post-graduate and doctoral degrees from the well-known technological institute – Veermata Jijabai Technological Institute (V.J.T.I.), Mumbai (formerly known as Victoria Jubilee Technical Institute). He graduated from this institute in 1970 and went on to serve a well-known textile group—Mafatlal Mills. After serving for nearly 5 years in the textile mill, he returned to his alma mater in 1975 as a lecturer in textile technology. In due course, he was promoted to assistant professor and professor.

In the last phase of his service at V.J.T.I., Mumbai, he took over as Head of the Textile Manufacture's Department. Almost during the same tenure, he held a position as Deputy Director in the same Institute. He has written several articles on card cleaning efficiency and the role of uni-comb, and carried out extended research on the influence of doubling parameters on properties of blended doubles yarns.

1 Carding

1.1 OBJECTS OF CARDING

Though a function of the blow room, a process prior to carding, is to open the matted mass of the fibres received from the bales, the process never reaches fibre-to-fibre separation. This again is true even with the modern blow room line. The main focus in this earlier operation has been to open out the baled cotton to the smallest possible size of the tufts and to carry out cleaning effectively. Therefore, it is the card which has to complete the action of fibre-to-fibre separation.

Apart from fibre individualization, the next important task that carding has to perform is to complete the cleaning action left over by the blow room. Many contrivances are added to carry out this objective in modern carding machines. Thus, whereas while the cleaning efficiency levels of earlier conventional blow room and card, when taken together, were in the region of 80–88%, the modern sequence reaches a much higher cleaning efficiency level between 95% and 98%. This is possible owing to a totally new approach of cleaning the material when it is being opened. Here the opening itself is intensified and made more effective.

In spite of the fact that the modern blow room uses new types of beaters that are equipped with finer spikes or saw teeth and focuses its attention mainly on producing very small tufts, the material that exits the blow room still does not reach the state of complete fibre to fibre separation. Similarly, even when a greater proportion of trash is extracted in the blow room, the unlocking of very small tufts in carding is mainly responsible for further releasing finer trash. This almost completes the cleaning.

With very effective seed traps used in the modern blow room, the broken seed particles are hardly allowed to go to card. In general, the material delivered by the card contains a very negligible proportion of trash. The sliver from a modern card is thus much cleaner. Such a clean sliver (< 0.2% trash) is able to meet the demands of new spinning technologies like Rotor, Dref or Air Jet spinning. In fact, the latest modern card is so well-equipped that, it has surpassed almost all the advantages of Tandem card* which was an intermediate stage between conventional cards and modern cards. The features of this technology are discussed later in this book.

Apart from cleaning and individualization, the card performs one more important function: removal of neps. The neps are tiny knot-like structures of highly entangled mass of fibres and they are very difficult to remove. The neps are usually formed due to the rolling action of fibres. This rolling may occur when the fibres rub against each other or when they pass either over the machine surface or during their journey through transporting pipes.

The removal of neps is possible in card, and this is basically in the region of cylinder and flats. However, it may be noted that wires in bad condition on both cylinder and flat are also responsible for the formation of neps. Thus, the

---

* Tandem card—Two carding machines suitably combined to improve its functioning.

DOI: 10.1201/9780429486562-1

excessive nep formation in the card is due to deterioration of card wires. This is because, when the wires lose their sharpness, they allow the fibre material to roll over their surfaces rather than precisely carrying it around. With more vegetable-originated impurities and a higher immature fibre percentage, this tendency of fibre-rolling has been observed to increase. In fact, the presence of neps in the card sliver web is one of the indicators of the quality of carding. The various features incorporated in modern carding aim at producing almost a nep-free yarn. If, however, some of the neps are left over, in spite of the best carding conditions, they would continue their journey up to the yarn stage and appear more prominently on its surface. Further, this would mar the appearance of the yarn, especially when it is dyed later. This is because these neps appear as tiny specks, owing to more dye absorption.

It is known that neps are removed in combing. The combed yarns, in comparison to carded yarns, therefore, are more nep-free. However, it may also be noted that the nep removal in combing is only incidental. This is because the basic object of combing is to fractionate the short fibres. While carrying out this fractionation, the neps are also segregated.

In the card, the transformation of the lap or loose fibrous material (in the case of chute feed) into a sliver thus involves not only great reduction in the size and thickness of material but also fibre to fibre separation.

The thinning-out operation, therefore, requires a very high draft to be employed in the card. Usually, this draft is around 100 and its value is kept slightly flexible so that a required hank of sliver can be produced after condensing the web delivered by the card. The typical arrangement of fibres in the card is discussed later in this book. The web condensed into sliver is subsequently coiled into a card can for its further processing.

The objects of the carding can thus be summarized as (1) fibre to fibre separation—*individualization*, (2) cleaning the foreign matter left over by the blow room, (3) removing neps and short, broken, immature fibres and (4) converting a lap or loose fibrous material from chute feed into a sliver. With this, it would be appropriate to learn some of the basic operations of the various elements in carding.

1.2 PASSAGE OF MATERIAL THROUGH CONVENTIONAL CARDING[1,2]

The lap produced in the blow room (in the case of chute feed, loose fluffy sheet directly entering the feeding system) is placed on the lap roller and a sheet of cotton is unrolled onto a horizontal table (Figure 1.1).

This table is shaped at its front end to form a feed dish plate (or simply feed plate). This typical shape accommodates a small feed roller. The material is thus guided through the feed roller and feed plate.

To control and grip the lap firmly, the feed roller is heavily loaded on either side. With this pressure, it precisely guides the forward movement of lap or material. It is very important that the feed roller grips the material firmly during its delivery; otherwise, the lap pieces or the loose materials are irregularly plucked by the subsequent saw-tooth covered roller Licker-in.

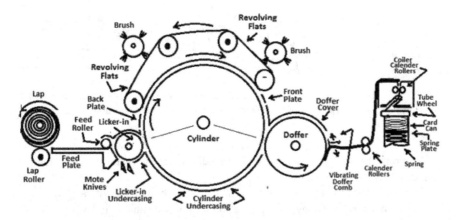

FIGURE 1.1 Passage of cotton through conventional card[1]: the specialized organs are involved in transforming lap (or chute-fed cotton) into a thick rope-like structure called *sliver.*

As the fleece passes through the feed roller and feed plate, it curves over the nose end, which is specially shaped so as to grip the lap until it is released. The lap is then subjected to an action of a fast-moving, saw-tooth covered licker-in roller (also called Taker-in).

The sharp saw teeth of the licker-in plunge into the fleece issued by the feed roller. The striking distance between the closest point of the feed plate and the points on licker-in saw tooth wires is very small (0.25 mm or 0.010 in). Each tiny wire point on the licker-in picks up very small quantities of cotton fibres and this helps in carrying out a very effective opening action. This releases a lot of imprisoned impurities and trash.

The licker-in works with a set of two mote knives. They are placed just below the feed plate and are positioned slightly tangential. They help in providing a better striking action and thus in jerking out the trash particles. For this, the mote knives have sharp edges, and when the small cotton tufts are struck by the licker-in saw tooth points, they help in loosening the trash trapped in these tufts. The licker-in undercasing helps in controlling the air currents generated by the fast moving licker-in and assists in preferential selection of lint. Thus, with the help of both the mote knives and the undercasing, a substantial amount of trash is extracted under the licker-in and, at the same time, the cleaned cotton lint is saved.

As the fibres are carried and guided by the licker-in wire points, they approach the cylinder, which is covered with much finer wire points of similar type (Figure 1.2a). The cylinder is considerably larger in diameter than the licker-in [50 in (127 cm) as compared to 9 in (approx. 23 cm)] and has almost twice the surface speed (660 m/ min) than the licker-in (320 m/min). The wire points of cylinder and licker-in, at their closest distance, are merely 0.127 mm (0.005 in) apart. It is interesting to note that, when in close proximity, the wire points of both are pointed in the same direction. All these factors enable an easy transfer of material from licker-in onto the cylinder. Thus, the cylinder wires merely strip the fibrous material from the licker-in surface. Further, owing to high surface speed, the cylinder carries around it air currents,

FIGURE 1.2   Transfer of fibres[2,3]: Four different modes of transfer of fibres from wire to wire. However, their purpose is quite different: (a) transfer, (b) condensation, (c) and (d) carding action.

which help in this stripping action. They also distribute the material across the cylinder surface.

After the material is transferred onto the cylinder, the wires on it carry the fibre-mass through a narrow passage—a gap between back plate and cylinder. This narrow passage and a steady distance-gradient between the entry and exit point helps in reducing the turbulence of wayward air currents carried around by the cylinder. In fact, it helps in streamlining them around the cylinder, thus helping uniform distribution of the fibre-tufts across the whole width of the cylinder and over its entire surface.

The fibres carried by the cylinder wire points then enter the cylinder-flat zone—the carding zone (Figure 1.2c and d). There are totally 105–110 flats, out of which, at any time, about 45–48 flats are in working position over the cylinder surface. The flats also have numerous tiny fine wire points on their surfaces and move in the same direction as that of the cylinder. However, as compared to the cylinder surface speed, they move at almost a snail's pace (10 cm/min). The wire points of the flats, however, are pointed in the opposite direction both to the cylinder rotation and to their own movement. Thus, the flats oppose the free flow of the cotton fibres around the cylinder-flat region. The distance between cylinder and flats, for each flat, is gradually closed down from 0.75 mm (approx. 0.03 in) at the entrance to 0.25 mm (approx. 0.01 in) at the exit. Thus, the fibres enter the gap between each flat and the cylinder with a wider distance at entry point, whereas they are smoothly pulled through a gradually narrowed setting as they come out of each flat. This induces a perfect carding action and it continues from one flat to the next. Thus, the carding action is carried over all the working flats. Thereafter, the fibres are carried further around the cylinder.

As regards the flats, they move slowly in the same direction over their working area on the cylinder surface, and finally move away from it. In due course, as they curve around and move away from the cylinder, they are stripped and cleaned by a very slow, oscillating flat wire comb and subsequently by a rotating brush with its bristles penetrating into the flat wires. This completes the work of flat cleaning. Thus, when the flats enter again from the back side (Licker-in side) into the carding zone, they are freshly presented for a continued carding action.

There is a constant exchange of cotton fibres between the cylinder and flat wires. Thus, the fibres carried by the cylinder wires are continuously transferred onto flats and back onto the cylinder. This results in a very effective opening of fibres and almost leads to a state of individualization. The action continues over the entire working portion of the flats. When the flats leave the cylinder, they hold a small percentage of fibrous matter—flat strip. The strip, when analyzed, contains the fibres deeply embedded in the flat wires, together with neps and fragments of vegetable matter. It is found that the mean length of the fibres in the flat strip, though less than that in the lap fed to card, does not differ significantly from it. On an average, the fibrous matter embedded in the flat wires constitutes a greater proportion of lint compared with licker-in droppings. Therefore, while the licker-in droppings appear to be far more dirty, the flat strip is more whitish in its appearance, and some of the fibres embedded in it are quite long.

The fibres on the cylinder continue their journey as they pass through the gap between front plate and the cylinder. In this context, an important function of the front plate can be realized. When the fibres are about to leave the cylinder-flat region, the front plate is positioned very close to the cylinder at a point where flats start coming away from the cylinder surface. The peculiar way in which the front plate is set, along with its position, induces the rushing air current from outside. The air currents try to rush-in through the gap between the back edge of the front plate and cylinder. During their journey, the air currents take some of the fibres that are long enough and are loosely held away from the flats back onto the cylinder. The front plate thus controls the flat strip.

The fibres carried further by the cylinder then come around and are closer to a slow-moving doffer. The doffer is also covered with wire points (Figure 1.2b) and is set very close to the cylinder at a distance of 0.125 mm (0.004 in). The sole purpose of the doffer is to take away the fibres which have undergone carding action between cylinder and flat, and carry them around for final delivery. The transfer of fibres is, no doubt, the result of the sudden release of air currents carried by the cylinder. The centrifugal force, the direction of the wire points and the curvature of doffer also help in transferring the fibres from the cylinder onto the doffer. However, all the fibres on the cylinder, as they come near the doffer wires, do not get transferred immediately in the same revolution of the cylinder. In fact, some of them may journey around the cylinder a number of times before finally being transferred. It has been observed that, on an average, some of the fibres may go around the cylinder as many as 18–20 times before they are finally picked-up by the doffer.

The distribution of fibres on cylinder wire points is in the form of a very thin layer. However, as a result of condensation, a somewhat thick layer is deposited on the doffer, and it is called as *card web*. This is because the doffer runs at a much slower speed (its surface speed varying from 22 m/s to 33 m/s). The level of condensation varies from 20 to 30 times. This fibrous matter is subsequently peeled-off by a fast-moving doffer comb (in high production carding, roller doffing is used; see Chapter 5, section 5.2.1) in the form of a filmy, semi-transparent web. The arrangement of the fibres in the web is most random and fibres are arranged in a crisscross manner. In fact, this very randomness of fibres in the web gives enough strength to the web to hold-on to itself against its own weight. A high degree of condensation on the doffer thus plays an important role in adequately strengthening the card web.

The doffer comb oscillations are required to be sufficiently quick so as to strip the web from the doffer effectively. Hence, it is necessary to adjust its speed in relation to the speed of the doffer. The construction and the mechanism of the doffer comb, however, limit the increase in its speed beyond a certain point.

The fibres peeled-off by the doffer comb and in the form of a web are condensed and calendered to get a compact, rope-like form called *sliver*. The sliver is further condensed and passed through a coiler tube to coil it into a card can. The weight of the sliver that can be accommodated in the can depends on the diameter and the height of the can. Many years ago these cans were simply dragged over the floor by the can carriers. With larger cans, as more material is coiled into it, they become quite heavy. Therefore, transport trolleys capable of holding several cans are used. However, still a much bigger can (42 in or 105 cm in diameter), when full, becomes far more heavy, and as such, cannot be simply lifted on to the trolleys for carrying it to the next process. In most of the modern spinning installations, very large cans (36 in × 42 in or even 42 in × 42 in) are used where the cans themselves are provided with rolling casters and can then be simply slid over the smooth floor. In such cases, the planning of the machinery is such that, with the blow room being placed into a separate room, the connecting pipes are run to directly feed the chutes supplying the loose and opened cotton to the modern card installations. Thereafter, the material needs some kind of transportation system to the next process. In a modern set-up, the card and the subsequent processing machine–draw frame are kept quite close, in the same department and under the same roof. In such a case (or otherwise also), the cans are provided with casters/rollers. It also becomes necessary to make the flooring very smooth (chemoxy flooring, magnesium oxychloride flooring, epoxy floor coating, etc.). This enables a smooth transportation of the card sliver cans (can trolleys or cans with casters) to draw frames.

REFERENCES

1. Cotton Spinning – William Taggart
2. Manual of Cotton Spinning – "Carding" – W.G. Byerley, J.T. Buckley, W. Miller, G.H. Jolley, G. Battersby & F. Charnley, Textile Institute, Butterworth publication, Vol. 3, 1965, Manchester
3. Elements of Cotton Spinning – Carding & Drawing – Dr. A.R. Khare, Sai Publication, 1999, Mumbai

2 Important Regions in Conventional Carding

2.1 FEED PLATE

A feed plate is also known as a *feed dish plate* as it is in the form of a rectangular dish (Figure 2.1)[1,2]. The lap roller, while rotating, unwinds the lap sheet at the rate of 25–28 cm/min. The sheet initially rolls over a plain surface of feed plate, which has a slight curved shape at its front end. A very small flat portion thereafter is called the *nose plateau* (Figure 2.1).

The initial curved shape of the nose conforms closely to the contour of the feed roller. It gives a greater and extended gripping action on the lap. Thus, the feed plate and the feed roller together firmly grip the entire width of the lap as it passes between them. The nose of the feed plate has a narrow horizontal plateau (Figure 2.2). The plate then is sharply bevelled towards the licker-in side.

This creates a wedge-shaped space (Figure 2.3) between the slanting bevelled face of the nose and licker-in wire points. The shape helps in providing a progressively intensified action of the licker-in wire points on the lap fringe. This is because the rate at which the fibres are fed to the licker-in is relatively slow (30 cm/min), whereas, once picked-up by licker-in wire points, the fibre speed is suddenly increased.

The fibres thus experience a very large accelerating force, which is likely to put a considerable strain on them. Therefore, owing to the wedge shaped space, the licker-in wire points gradually but steadily penetrate into the lap fringe. Thus, the suddenness of the action and its severity are both very much reduced. Even then, the effectiveness of the licker-in action in opening the tufts is never reduced. A very close setting is maintained between the feed plate and licker-in at the point B. (Figure 2.4). The normal setting is 0.25 mm, though it may be slightly changed depending upon the staple length. It may be pointed out that without the wedge-shaped space, the fibres would have been simply pulled and suddenly made to pass through this narrow setting distance.

The distance between the two points on the feed plate—point A, where the lap is held by the feed roller over the feed plate and where it is ultimately released at point B, which is the nearest point of approach of licker-in wires (Figure 2.5)—should be slightly greater than the staple length of the fibres processed.

This is because if the fibre length is longer than this distance, the fibres gripped at A would experience a severe strain at their other ends. As a result, the fibres, under this condition, are likely to be either damaged or broken. It may also be noted that the gradual penetration of licker-in wires into the lap fringe, from point A to point B, does not reduce the ultimate fibre opening action of the licker-in; it only reduces its suddenness.

DOI: 10.1201/9780429486562-2

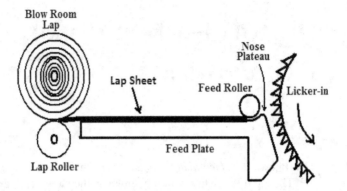

FIGURE 2.1 Position of feed plate[1,2]: The plate supports the lap between lap roller and licker-in, and prevents its folding.

In the past, feed plates with varying tapered lengths (distance AB—Figure 2.5) were used. It is obvious that a longer length of this tapered face of nose would provide a longer distance AB. This is done to suit processing of longer-staple fibres.

However, in the normal day-to-day working of a mill, it is neither possible nor essential to keep sufficient stock of such feed plates of different sizes. It would unnecessarily increase expenditure on spares. Therefore, in the absence of such appropriate feed plate/s, the effect is still produced by correspondingly widening the setting between the licker-in and the point B (setting B'—Figure 2.6). This, in a way, increases the distance A–B. Hence, when processing a longer staple material, a wider setting at B is used.

2.2 FEED ROLLER

The feed roller has flutes on its surface. The flutes help in exercising adequate grip on the lap fringe passing through the roller and feed dish plate. The roller itself is held at its two ends by bearing blocks which ensure its smooth running. The blocks, in turn, are held by brackets carried by the feed plate (Figure 2.7). The feed roller is kept almost flushing with feed plate throughout its length so that the whole length of lap is not only held firmly but also uniformly across the width. This ensures that none of the lap portions are allowed to be prematurely plucked by moving licker-in teeth.

Additional weights are put on the bearing blocks on either side, so as to increase feed roller loading. In earlier systems, the feed rollers were weighted by hanging dead

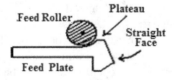

FIGURE 2.2 Nose plateau of feed plate[1,2]: It is specially designed to enable the perfect control over the lap fed to licker-in.

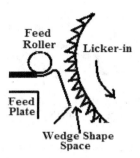

FIGURE 2.3 Wedge shaped space[1,2]: This provides gradual penetration of licker-in teeth into the lap fringe.

weights; however, the later makes incorporated lever weighting systems (Figures 2.8 and 2.9). In a modern card, they are pneumatically weighted. The feed rollers should exert uniform pressure on the lap fringe issued by them. This can be checked by putting a piece of paper between the roller and feed plate. If the paper is uniformly gripped across the width, the nip can be assumed to be uniform.

Yet another way of checking this would be to observe the front edge of the lap itself. This is done by disconnecting the feed roller driving gear (in old cards, it was the *change pinion*). The lap is then unrolled to inspect its front edge. In this case, the front edge of the lap should be uniform. Any irregularity in the front edge reflects undue snatching of the lap front edge. This indicates inadequate gripping of the lap by the feed roller at that point. When the card is stopped for setting, it is customary to take the feed roller carefully out of its bearing-blocks. During this time, any embedded matter in its flutes may be removed. It is also advisable to polish the flutes so as to make them sharper and more effective, thus improving their grip. The cleaning and lubrication of the block-bearings can also be done during the same period.

While putting the feed roller back in its position, the bearing blocks holding it on either side should go freely into the slots of the bearing bracket. Care must be taken to see that, with the lap present under the feed roller, the bottom part of the bearing block should be at a small distance from the top face of the bearing

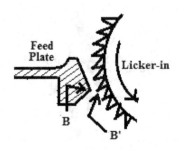

FIGURE 2.4 Setting of feed plate[1,2]: The feed plate is set to the licker-in at its closest point. This automatically forms a wedge-shape above.

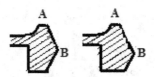

FIGURE 2.5   Feed plates[2,3]: Different types of plates were available in the past and they provided different distances AB to suit fibre length.

bracket (Figure 2.10). This gap ensures that when the bearing block is loaded with additional weights, the pressure is fully exercised and is made to act on the lap being issued.

The teeth on the change pinion and the plate wheel (in old card) must also be checked occasionally for any damage to their teeth; otherwise the drive to the feed roller will not be smooth and continuous. When this is ensured, it takes care of even and uniform feeding to the licker-in. Any uneven feeding is likely to endanger the fibres as well as the licker-in wires.

2.3   LICKER-IN[2]

It is a shell roller with a diameter of 23 cm (9 in") and has a series of uniformly spiral-cut grooves (Figure 2.11a). There are 6–8 grooves/ in along the length of the licker-in and they accommodate the base of the licker-in saw-tooth wire. When the wires are thus fixed into these grooves, their bases (or ribs) are totally buried inside, whereas the full teeth project out of the licker-in shell (Figure 2.11b). The licker-in, with these grooves, looks like 6 or 8 start worm. Thus, when the licker-in wire is being mounted, one wire mounting is started from end A (Figure 2.11c). The mounting continues beyond point 6 (where six starts have been completed) and finally reaches the other end.

Depending on the number of starter-grooves (6 or 8) around the circumference at end A, separate wire-starts must be mounted in respective grooves until all the licker-in is fully clothed. Thus, the total mountings will require repeating the operation six or eight times for a fully clothed licker-in. There are approximately 30,000 to 40,000 fine saw tooth points, much like a series of closely spaced circular saws,

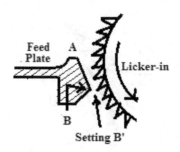

FIGURE 2.6   Setting of feed plate[1,2]: A carefully carried-out setting ensures gradual penetration of licker-in wires and controls fibre damage/breakage.

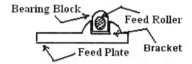

FIGURE 2.7 Bearing blocks[1,2]: They hold the feed roller shaft on either side.

on the licker-in. With them, the licker-in is able to greatly reduce the size of the tufts fed from the feed system.

However, it may be mentioned that the licker-in is not capable of individualizing the fibres. The very tiny clusters picked-up by the licker-in teeth easily contain about 8–10 fibres; the degree of reduction in the size of fibre clusters is quite effective in bringing about the required cleaning, though.

For treating different classes of cottons, the licker-in action can be varied by changing its speed, the number of teeth per inch on the licker-in wire and their angle of rake. Licker-in wires having a varying number of teeth per unit length are available. Thus, wires with a minimum of 3 teeth/ in up to any suitable number higher than this are commercially used for different degrees of licker-in action. For coarser and trashy cottons, a stronger licker-in action is required and hence teeth per inch on the wire are far more than this minimum. The longer staple cotton, on the other hand, is comparatively delicate and clean, and the number of teeth per inch on the licker-in wire is notably less.

Even the speed of the licker-in can be varied depending on the class of cotton. A wide range of licker-in speeds is in commercial use. Thus, on a conventional card, for processing long staple and fine cottons, comparatively lower speeds from 350 to 420 rpm are used; whereas for very short and trashy short staple cottons, speeds beyond 660 rpm are popular. As mentioned earlier, the licker-in speeds must be related to cylinder speeds. Therefore, when licker-in speeds are increased, it is necessary to see that the cylinder speed is increased proportionately to maintain the ratio of their surface speeds. The increased licker-in speed should not, in any case, bring this ratio (cylinder to licker-in) lower than 1.85–1.9.

On a normal high production card, much higher licker-in speeds are used (830–880 rpm). In the latest versions of modern card, still higher speeds (around 1040

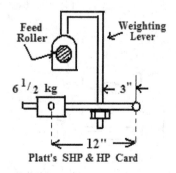

FIGURE 2.8 Feed roller weighting.

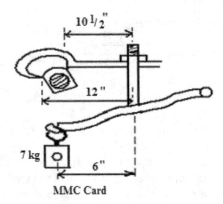

FIGURE 2.9 Feed roller weighting.

rpm) are very common. With the higher licker-in speed, more wire points pass through the lap fringe, and consequently, its opening action is far more powerful. However, the momentum with which the points plunge into lap fringe is also very high. Though this leads to very powerful opening action, it is likely to cause certain damage to the fibres. This is precisely why lower licker-in speeds are used for longer staple fibres, as these fibres are held under the feed roller for a longer time. Further, the longer fibres are comparatively more delicate and also contain a much lower percentage of trash. They, therefore, need a gentler treatment in the licker-in zone.

In addition to this, the angle of inclination of the licker-in wire teeth has a profound influence on their action on the fibres. The forward (positive) rake (Figure 2.12) wire is generally used for coarser mixings when a somewhat harsher treatment is required. With the teeth bent in a forward direction, the fibres are held more positively by the wire points. The perpendicular or 90° rake is mostly used for finer varieties of cotton where the trash content is notably lower. With a 90° rake, the treatment given to the fibres is comparatively mild (Figure 2.13). Because the front side of leading edge of the tooth is upright, the fibres are held less tenaciously, and as such, are released onto the cylinder more easily. The negative rake, on the other hand, is exclusively used for synthetic and man-made fibres (Figure 2.14). These wires have the least holding

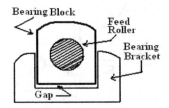

FIGURE 2.10   Feed roller bearing block[1,2]: It holds the feed roller with its bearings. The small gap between the bearing bracket and the block ensures full loading on feed roller.

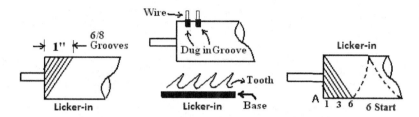

FIGURE 2.11   (a)—(c) Grooves on licker-in[1,2]: Spirally cut grooves on bare licker-in surface in a way, decide the point density of wire points. (a) Groove spacing, (b) dug-in grooves and (c) starts around the licker-in circumference.

power and mainly open the tufts. This is because there are no impurities in terms of trash in these bales.

The purpose of licker-in action, in this case, is just to open these fibre tufts for the cylinder wires to complete the individualization process. The teeth are slightly inclined in a backward direction, so the fibres are easily released by the wires when they approach the cylinder. Consequently, the quick fibre transfer onto the cylinder avoids any possibility of overbeating the man-made fibres, and yet effectively opens-out the fibre clusters.

2.3.1   TYPES OF LICKER-IN WIRES[1,2]

The licker-in wire has to be very strong so that it is able to penetrate the compact mass of lap fringe. The wires, therefore, are made of special steel with hardened and tempered points. These wires are typed commercially for their use and are accordingly labelled in different ways for the purpose of ordering. Apart from teeth per inch, the height, the angle of rake, and point density, the licker-in opening action, as mentioned earlier, can be varied by changing its speed. The effect of point density and speed is realized together. This is because each point is responsible for taking away a small, tiny tuft of fibres from the lap issued by the feed roller. Therefore, the point density together with the licker-in speed decides the number of points available per unit time to penetrate the lap fringe. The base length of the tooth along the wire decides the pitch (teeth per inch), which again is related to point density. In the case of a grooved licker-in, it is the rows per inch (along the length of the licker-in) and the pitch that decide the point density.

As mentioned earlier, the *carding angle* is very important for card wires. It decides the intensity or aggressiveness of the action of the wire teeth. Earlier, it

FIGURE 2.12   Forward rake. Forward rake angle 73° for coarse and medium cottons.

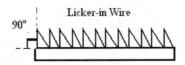

FIGURE 2.13 Perpendicular rake. Perpendicular rake angle 90° for finer varieties.

was defined as the angle of the leading edge with horizontal. In present techno-logical terms, this angle is defined as the angle of the leading edge to the vertical.

For a licker-in wire, it varies from +20° (70° from horizontal) to –10° (negative rake for human-made fibres).

For the cylinder, it varies from +10° to +20° (70° to 80° from horizontal), and for the doffer, from +25° to +30° (60° to 65° from horizontal).

Knowing the licker-in (or cylinder or doffer) circumference and its length across the carding machine, the point density (i.e., points per unit area) can be easily cal-culated. The wire particulars required in this case, are: (1) base or rib width (for cylinder doffer and number of start-ups for licker-in), and (2) teeth per unit length of the concerned wire.

Consider a distance of 1 cm along the length of the licker-in. With a base width (wire rib-width) 'r' cm, there would be 1/r coils per cm around the licker-in. Further, if there are 't' points per cm along the length of this wire, then 1/r coils will have $(1/r \times t)$ points over the surface area of 1 cm².

Thus,

$$\text{Points per square centimetre} = \frac{t}{r} \tag{2.1}$$

As $t = 1/\text{pitch}$, we have,

$$\text{Points per square centimetre} = \frac{1}{\text{Base width cm} \times \text{Pitch cm}} \tag{2.2}$$

The above formula for an area of 1 in² can be easily expressed, knowing that

$$1\,\text{in}^2 = 2.54 \times 2.54 \text{ cm}^2$$

Thus, Or Points per square inch $= \dfrac{645}{\text{Base width mm} \times \text{Pitch mm}}$ (Note: It is mm)

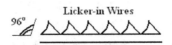

FIGURE 2.14 Negative rake. Negative rake angle (–5° to –10°) for man-made fibres.

TABLE 2.1

Range of Licker-in Wire Particulars

Height	Rib Width	Teeth/in	Angle from Vertical	Point Density
5.0–5.5 mm	1.1–1.5 mm	3–5 normal	0°–20°	35–205
		4–13 interlock	5°	

The normal range of specifications for licker-in wires is given in Table 2.1. The card wires are manufactured by several companies, viz. Indian Card Clothing (ICC), Trutzschler Card Clothing (TCC), Lakshmi Card Clothing (LCC), Graf, Uni-spin Card Clothing (UCC), etc. normally fall within the above range.

Basically, licker-in wires are of two types—grooved and interlinked (interlock). Conventionally, the wires are forced into the grooves dug on the licker-in surface.

The interlinked wires are made to link and hence merge with each other. They remain on the surface of the licker-in.

LCC-Ultima and Chroma (Lakshmi Card Clothing) have introduced an interlock type of wire made of steel. It can be mounted into V-shaped groove. Graf uses Cutty-Sharp steel alloy for its licker-in wire.

A standard licker-in wire is used for cottons from 16 mm to 25 mm staple; there are 1.8 teeth/cm (4.5 teeth/in) on the wire. The angle of leading edge of the tooth is +17° (from vertical) whereas the overall height is 6.3 mm (0.249 in).

A special wire is offered for cotton staple varying from 28 mm to 38 mm. The overall height of wire is slightly less than standard wire. However, the number of points is reduced to 1.4 teeth/cm (3.5 teeth/in). The angle of the leading edge is 0°, i.e., perfectly perpendicular. Consequently, for fibre lengths of 38 mm and above, linear teeth density is still further reduced to 1.25 teeth/cm. For synthetic and man-made fibres, the lowest teeth density of 1.2 teeth/cm (3 teeth/in) is used. In this case, the leading angle of the wire is negative (–6° to –8 from vertical, in the reversed direction).

Generally, with 6 to 8 starts (similar to a 6 or 8 start threaded worm), there may be 3 to 6 teeth/cm² (18–36 teeth/in²), on the clothed surface of the licker-in. Latest generation carding machines offer as many as 16 rows/in along the length of the licker-in. With this, it is possible to increase the point density up to 200. Figure 2.16 shows the typical wire profiles. The intensity of the licker-in wire is strongly related to the angle of the leading edge of the wire (Figure 2.17). The positive rake has powerful opening & cleaning action; whereas the negative rake acts comparatively softly

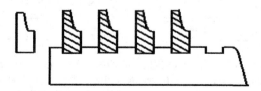

FIGURE 2.15 Grooved type.

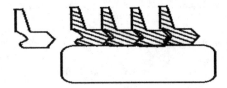

FIGURE 2.16   Interlinked type.

on the material and is preferred for all man-made fibres (viscose, polyester etc). The perpendicular rake is used for long staple cotton with very low trash content.

The helical winding of the wire, along with a high speed of the licker-in, makes its action very powerful and effective. The count of wire (thickness or gauge) is selected so that the wire is very strong. The wider spacing of the teeth also prevents clogging of fibres or impurities between the teeth.

Owing to the saw-tooth construction, the licker-in metallic wire is perhaps the strongest among all the types of metallic clothings used on the card. There is greater friction between the wires and fibres. The wider spacing for the teeth also gives an added advantage, in that the lap fringe is presented simultaneously to the leading edge of each tooth as well as its points. This almost completes the penetration of the saw-teeth into the lap fringe quite effectively and satisfactorily.

2.4   MOTE KNIVES[2]

Mote knives are knife-edged rectangular bars, approximately 3 inches in height and extend the full length of the licker-in. The knives are in pairs (Figures 2.18 and 2.19) and are fitted beneath the licker-in, in between the feed plate and the undercasing. The bevel shape and the knife edge give the mote knives a superior cleaning power. These knives can be adjusted independently to get the desired degree of closeness to the licker-in wire surface.

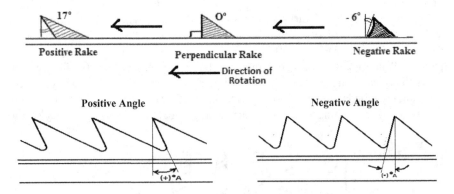

FIGURE 2.17   Angle of inclination of licker-in wires[1,4]: The intensity of action of wires on cotton depends on the carding angle—the angle of the leading edge. The positive rake gives very strong action whereas a negative rake gives comparatively milder action. The perpendicular rake lies between these two.

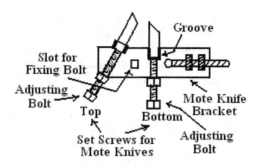

FIGURE 2.18   Mote knife bracket.

The mote knife bracket bears two grooves to carry the two knives, which fit snugly into these grooves and are firmly carried. Owing to this, the mote knives can be presented to the licker-in wire surface at a certain fixed angle. When two knives are used, the first or top knife is set closer to the licker-in surface: 0.254 mm (0.010 in). The second or bottom knife is set slightly farther away, at 0.38 mm (0.015 in).

When the small fibre tufts carried by the licker-in wire points hit the top mote knife almost tangentially, the force is maximum and very effective. Though the bottom mote knife also acts as a striking element, it also prevents the fibres from going further down along with the trash extracted at the licker-in. More than half of the trash present in the lap is extracted at the licker-in region, and mote knives are mainly responsible for this. Therefore, the setting of these two knives needs to be precisely carried-out. The mote knives, when set judiciously, help in dislodging loosely held impurities, which mainly constitute the licker-in droppings. However, it is possible that along with trash, some good fibres are also lost under the licker-in. If this happens, it may require reviewing the mote knife setting.

The setting of the top mote knife can be adjusted by moving the mote knife bracket closer to the licker-in along the horizontal plane. Once this position is arrived at, the bracket bolts are securely fastened. The setting of the bottom knife is then carried out by lifting or lowering the knife vertically in the slotted groove by a set screw provided under the slot. There are two such brackets, one each on either side, to hold the full length of the knives firmly. For more safety, some mills provide a third bracket at the centre. This gives additional support to the knives. With this bracket, there is less danger of the long length of knives buckling at the

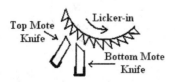

FIGURE 2.19   Positioning of mote knife.

centre due to sudden stresses developed, especially those due to a thicker portion of lap suddenly entering the licker-in zone. In the past, there had been a few serious accidents when, without central support, the mote knives yielded to such sudden stresses and broke. In certain instances, the broken knife portions were observed to be carried further by the licker-in up to cylinder, thus causing severe damage to its clothing.

The mote knives derive their name from their usual function. They remove motes very effectively. The motes are vegetable impurities, such as cotton stalks, leaves, seed particles, etc. In spite of cleaning carried out in the blow room, these motes still remain present in the lap. The mote knives thus play an important role in cleaning such impurities. The impurities thus extracted with the help of mote knives by the licker-in, drop down under it through the gap between (a) the feed plate and top knife, (b) the two knives and (c) the bottom knife and licker-in undercasing, and are called *licker-in droppings*.

2.5 LICKER-IN UNDERCASING[1,2]

The licker-in undercasing is a screen partly covered with a perforated sheet and, at the front, by 2 to 3 spaced grid bars (Figure 2.20). The back end of the undercasing is hinged with the cylinder undercasing, whereas its other end (which is towards the mote knives) is carried by the two links that are supported by the two brackets. These brackets are bolted to the inner side of the machine framing and are placed on either side.

The brackets are adjustable. The undercasing comprises three parts—nose, grid section and perforated screen. The nose is a smooth strip, rounded at the front. It is mainly responsible for guiding the air currents generated by the licker-in and keeping them close to the licker-in surface. The actual separation of lint and trash takes place when the air currents carrying lint and trash around the licker-in get bifurcated at the nose portion of the undercasing. According to aerodynamic principles, the lint, which is lighter, tries to remain closer to the licker-in surface, whereas the trash, which is heavier, remains in the outer layers of air currents.

A correct positioning of the nose is, therefore, very important. It keeps the lint close to the licker-in surface. As the nose is made to curve down (Figure 2.21), it diverts the path of the trash and leads it to move away and down the licker-in. It may

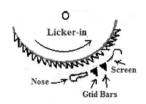

FIGURE 2.20 Licker-in undercasing[1,2]: It is basically a lint supporting sieve placed under the licker-in. The leading edge, called nose, diverts air currents to minimise lint loss.

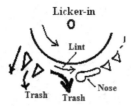

FIGURE 2.21   Licker-in.

be noted that the lint and the trash carried by the licker-in air currents never truly occupy two distinct layers. The lint and trash combination is richer in lint in the layers closer to the licker-in surface (Figure 2.22). The proportion of trash goes on increasing from inner to the outer layers of air currents around the licker-in.

Therefore, if the undercasing is set closer to the licker-in, it may divert most of the trash away from the licker-in and it would form licker-in droppings. However, these droppings would contain a sizable proportion of lint, as well. On the other hand, a wider setting of the undercasing nose may allow all possible lint to go through the gap between the licker-in and undercasing, but a significant proportion of trash would also go along with lint. In the second case, therefore, the cleaning efficiency of the card would be severely impaired.

Therefore, an optimum setting has to be found by trial and error. With this optimum setting, it would be possible to achieve maximum cleaning with minimum lint loss in the droppings. It is observed that around this optimum setting, the licker-in droppings usually contain about 30% lint and 70% trash.

Next to the nose is the grid section, which mainly provides small openings for finer trash to pass through and get extracted. The number of grid bars varies with the type of undercasing. The opening between the two adjacent grid bars varies from 4 mm–8 mm (5/32 in to 10/32 in); whereas the top face of the grid is approximately 4 mm wide. The shape of the bars is available in two types—M and N (as shown in Figure 2.23). The total distance from the nose to the last grid bar of the undercasing varies from 6.25 cm to 7.0 cm (2.5 into 2.75 in).

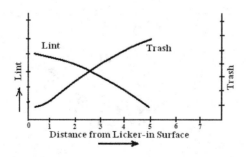

FIGURE 2.22   Lint and trash in air currents surrounding the licker-in.

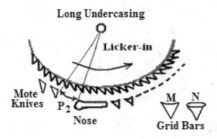

FIGURE 2.23   Short undercasing.

The perforated section, called *screen*, is mainly provided to remove a much finer, dusty type of trash in the lap fringe. The pore size of the perforations varies from 0.8 mm to 2.4 mm (1/32 into 3/32 in). The perforated area also partly serves to partially vent-off feeble air current. The overall length of the undercasing from nose to the end of the screen section varies from 20 cm to 28 cm (8 in to 11 in); accordingly, the undercasing is termed *short* or *long* respectively (Figures 2.23 and 2.24).

The short undercasing, when placed in its position, is at a greater distance (P_1 in Figure 2.23) from the bottom mote knife, whereas the longer undercasing, having longer reach, extends closer to the mote knives (P_2 in Figure 2.24). It is evident that with wider spacing between the mote knives and the undercasing (P_1), there is more opportunity for the trash to fall down, and consequently, this type of undercasing (short) is more useful for short staple and trashy cottons. Obviously, the longer undercasing providing a smaller gap (P_2) acts as a lint saver and therefore partially avoids lint loss in the droppings. The undercasing is set to the licker-in by a sweep gauge (Figure 2.25). It is customary to keep a certain gradient in setting the distance from nose to the end of the screen portion. Near the cylinder side, undercasing is set at 0.86 mm (34/1000 in); whereas at the nose, it is set at 3.1 mm (125/1000 in). In special cases, a still wider setting of 6.3 mm (¼ in) can be used at the nose, especially when processing longer staple cotton with lower trash content.

The purpose of keeping this kind of gradient in setting from nose to perforated screen is to streamline air currents around the licker-in. If the air becomes turbulent,

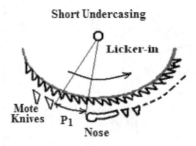

FIGURE 2.24   Long undercasing.

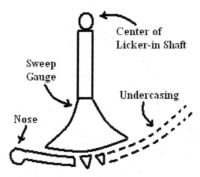

FIGURE 2.25 Undercasing setting[1,2]: Special care is taken to maintain a small gradient from entry point to the end of the undercasing. This helps in streamlining the air currents and thus saves lint in the dropping.

the back pressure developed at the nose would disturb the lint and trash separation. It is very important to periodically inspect the undercasing surface, especially the one facing the licker-in. It should be very smooth as this allows an easy sliding of lint. So also, the joining of the three sections—nose, grid and perforated screen—need periodic inspection. In some instances, it becomes necessary to solder the loose joints. Usually this is done during the *half-setting** procedure.

2.6 BACK PLATE

The sheet extending from licker-in cover to a point where flats begin working over the cylinder is called the *back plate*. It is secured to circular-cylinder bends on either side of the card. The upper and lower settings are carried-out with the help of two adjusting screws, A and B (Figure 2.26), respectively. These set screws rest on the bend, whereas another check bolt C fastens the back plate on the bend itself and acts as retaining screw. Thus, when setting the upper or bottom edge of the back plate, the check bolt S C needs to be loosened first. The setting at each end can then be changed with the help of adjusting screws A or B. Again, after the final setting is done, it is equally important to fasten check bolt C securely.

The lower edge of the back plate controls the air currents generated by the fast revolving cylinder. These air currents try to rush into the main carding zone and disturb the uniform distribution of material received from licker-in. A small gradient (from B to A) avoids any turbulence and streamlines these air currents.

With an ideal setting, the distribution of the fibres coming from the licker-in onto the cylinder surface is fairly even. This helps create a very effective carding action. Further, the control of air currents reduces any back pressure developed at the cylinder-licker-in junction and also at the nose of the licker-in undercasing.

---

* Half-setting—A periodic cleaning and setting on the licker-in side when the whole licker-in assembly is taken out for cleaning, oiling and greasing.

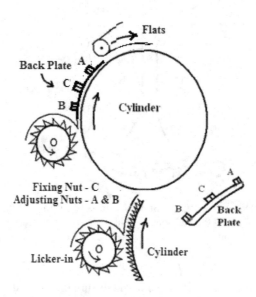

FIGURE 2.26 Back plate[1,2]: It controls and streamlines the air currents around the cylinder and helps in uniform distribution of fibres on it.

Thus, a uniform distribution of fibres on the cylinder leads to improved carding action. This also helps in reducing fibrous matter in the flat strip, which then becomes richer in trash content. The setting of the upper edge of the back plate (towards the flats) acts as a seal to limit the air currents. If this setting is too wide, the cotton is blown out between the flats. This disturbs the functioning of the flats, and in turn, affects the flat strip percentage. The setting of both the lower and upper edges of the back plate controls the appearance of card web delivered by the doffer. Too wide a setting at either edge may lead to *cloudy* (whitish) web due to insufficient carding action. The cloudy web portions are patches of inadequately carded material and can be easily noticed in the web delivered by the doffer.

2.7 CYLINDER[1,2]

The cylinder is primarily responsible for the real carding action. The carding as defined is –"To reduce the entangled mass of fibre-tufts into a fibre-to-fibre state (individualization)"—and to convert it into a thin filmy web. For this, it is necessary to work the fibres between the two closely spaced surfaces clothed with opposite sharp wire points and moving with large relative speed difference." A carding action (Figure 2.27) is thus obtained when the cylinder wires and flat wires work point to point, in close proximity, with opposing wire points and at high relative speed difference.

A much earlier stage in the carding involved the use of stationary flats (hand and Stock card). However, with the introduction of revolving flats, the carding becomes continuous and faster.

By the above definition of carding, there is no carding action between the feed roller and licker-in. There is only reduction of lap pieces into smaller clusters. Even the action

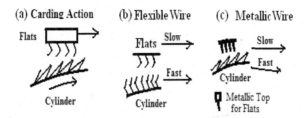

FIGURE 2.27 Carding action[1,4]: The true carding action takes placed only when the wire points on the two surfaces are acting in opposite directions; the two surfaces are placed in close proximity and there is a large relative speed difference between these two surfaces. (a) Carding action, (b) flexible wire, (c) metallic wire (carding action with opposite wire points).

between licker-in and cylinder is more of stripping nature. This is because the cylinder, owing to its faster surface speed, is merely able to carry the fibres from licker-in. In spite of opposite spikes, there is also no carding action between cylinder and doffer; on the contrary, there is a mere condensation of fibres on the wire points of the doffer. The real carding action, therefore, is only between the cylinder and the flats.

The main cylinder (Figure 2.28), which is 127 cm (50 in) in diameter, is called the *heart* of carding. It is primarily a hollow iron shell (in high production card it is made of steel) cast in one piece with strengthening ribs running inside. Each end of the shell is machined internally to take a cast iron cylinder-ending or *spider*. The spiders on the two sides provide spokes and hubs for complete cylinder formation. The cylinder shaft is made to pass through these hubs.

There are conical bushes (Figure 2.29) between the cylinder shaft and the hub on either side. The bushes are tapered inwardly and are inserted until the cylinder is correctly positioned and securely held on the shaft.

The cylinder shaft is supported near each end with thrust and radial bearings carried by the pedestal brackets. These brackets at the bottom are bolted and fastened to the top of side framings on either side.

This secures the position of the cylinder shaft exactly perpendicular to the length of the framing in the horizontal plane. It is essential that the cylinder, when placed on cylinder brackets, be correctly aligned, securely mounted and dynamically balanced.

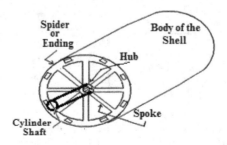

FIGURE 2.28 Bare cylinder[1,2]: Under the wires, the cylinder is a bare and smooth cylindrical body. It is made of strong steel to hold the heavy weight of cylinder wires and bear the pressure with which this wire is mounted on it.

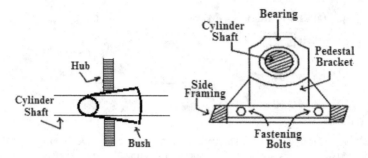

FIGURE 2.29   Cylinder supports[2]: Cylinder being very heavy needs strong and sturdy bearings to hold its shaft. These bearings must provide a firm seating and should not allow any vibrations to setup when the cylinder is made to rotate at high speed.

Equally important is ensure that the cylinder is properly machined and ground to give absolute concentricity of its surface at all points. On a modern card there are only metallic wires. However, it will be interesting to note that the flexible wires almost ruled the carding world till 1955. When preparing the cylinder for flexible wire mounting, the transverse rows of holes (Figure 2.30a) along the machine width were initially dug into its surface.

The tapered wooden bushes or plugs were then driven into these holes. By grinding the surface, these bushes were smoothened out later. With this, it was possible to perfectly align the top surface of the bare cylinder.

However, when the cylinder is originally clothed with metallic wire, all this procedure is not necessary. This is because the metallic wire is directly mounted on the bare cylinder without any foundation. When mounting the metallic wire, the winding is started from one end. In this case, it becomes essential to retain the first extreme wrap in its position. For this, grooves are cut on either side and at the end of cylinder surface so as to fit Z-wire (Figure 2.30b) to give support to the extreme end layers of metallic wire on the cylinder. The ends on both sides of the cylinder are closed by side metal sheets to prevent undesirable air currents from disturbing its functioning during working. These metal sheets are secured to the end rings or spiders on either side by means of counter-sunk screws. The pulleys on either side

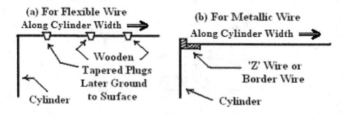

FIGURE 2.30   Mounting of wires[2]: The flexible wire with a firm foundation requires a perfect seating on the bare cylinder. The provision is made to hold the foundation by inserting screws through the foundation into the wooden plugs held by the cylinder surface. Metallic wire however, is simply wound on the levelled surface of the bare cylinder. (a) For flexible wire along cylinder width and (b) for metallic wire along cylinder width.

of the shafts are then put. The bare cylinder is finally given a light surface grinding prior to wire mounting. It is at this stage that the cylinder is given a thorough checking for both static and dynamic balancing.

2.8   FLEXIBLE WIRE POINTS[1,2]

Before the introduction of metallic wire, flexible wire was used for covering both the cylinder and the doffer. In this case, the wire clothing was constructed by inserting fixed lengths of flexible wires into the backing material or *foundation*.

Each wire piece, in the form of two limbs of a U, was stapled into the foundation. Thus, at the back of this foundation, a crown was formed, whereas on the face, the two legs of the wire, pierced through the foundation, appeared with two sharp points.

The wire, as it emerged out on the face, was also given a typical bend for its upper portion. It is called the *knee* (Figure 2.31). The formation of this knee was brought about during piercing itself by bending the two legs over a bar before the full crown was forced through the foundation.

This helped in maintaining consistency in both the height and the angle through which the wire was bent. It was possible to vary the angle at which the wires were forced through the foundation, the angle at the knee, the height of the knee above the foundation and the length of the wire above the knee. The overall height of the wire from the crown to the wire tip was 9.52 mm (0.375 in), whereas the total length of the wire from one point to the other for the same crown was 22.19 mm (0.875 in).

Typical flexible wire clothing was made on a specially designed automatic setting machine. The rolls of the wires, as well as the foundation material, were fed to the machine, which worked at a speed of 400 staples/min. A typical construction of cylinder wire had 8 staples in a row and approximately 24 rows/cm (60 rows/in) along its length. The length of the wire fillet for a fully clothed cylinder varied from 75 m to 90 m, whereas that for doffer varied from 30 m to 45 m.

2.8.1   FOUNDATION MATERIAL FOR FLEXIBLE WIRES[1,2]

The object of the foundation was to hold the wires securely in their position. For preparing the clothing for cylinder or doffer, the foundation materially, originally of much wider width, was used for inserting wire staples. Later, it was cut to suitable width, for ease in mounting. The common width for cylinder fillet was 2 in and that for doffer fillet was 1½ in.

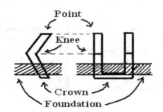

FIGURE 2.31   Flexible wire[1,2]: The typical feature of flexible wire is the bent knee. It is that portion of the wire above which there is a sharp point. This point brings out carding action.

Much earlier, leather was used as the foundation material. It is still used in woollen carding. Leather is very strong, rigid, and to a certain extent resistant to moisture and oil. However, it lacks in elasticity. The most popular foundation material was made by bonding cotton-linen fabrics by rubber solution. Another alternative was the cotton and wool sandwich foundation, bonded using glue and oil cement. The plied-fabric foundation became very popular. It was not only rigid and strong but also comparatively cheap. The structure was very uniform and could be varied by changing the plies to suit the requirements.

The standard CWC foundation had 3 plies: (a) cotton face, (b) linen warp and woollen weft at the centre, and (c) cotton-backed cloth. It was usually bonded with glue and oil. The rubber was sometimes used to give pliability to the structure. Along the same lines, four-ply CCWC was produced. A well-known 5-ply VIR was made by adding cotton ply and then using a vulcanized rubber face.

It is essential that the foundation material be selected to suit the climatic condition. It must be sufficiently strong and at the same time flexible enough to permit its application on the circular surfaces of both cylinder and doffer. It must not stretch, or become loosened or blistered. It must be rigid enough to hold the wires in their position, but at the same time allow resiliency for a slight displacement when working under the pressure. When the pressure is released, it is essential that the wires return to their normal positions. Finally, the foundation material was expected to be resistant to moisture, oil, heat and premature ageing.

2.8.2 Setting Pattern of the Crowns in Flexible Wire Clothing[1]

In setting the wire crowns, different designs, like plain, twill or sateen were used. In Figure 2.32, each pair of circles represents the two wire points, whereas the line joining them is the crown. This crown arrangement is seen from back of the foundation clothing.

There are approximately 50 to 80 rows of crowns per inch along the length of wire clothing. Most commonly used sets are rib for the cylinder and twill for the doffer;

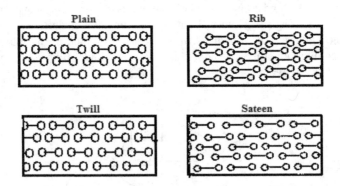

FIGURE 2.32 Pattern of crowns[1,2]: The crowns are formed at the back of the foundation with a pattern based on basic weaves used in fabric manufacture. In a way, they decide the openness or closeness of points.

TABLE 2.2

Count, Gauge and Points per Square Inch

Count of Wire	Gauge	Diameter (mm)	Points/cm²
90	0	0.36	69
100	31	0.33	77
110	32	0.30	85
120	33	0.28	93
130	34	0.25	100
140	35	0.23	108

however, the plain set can give maximum point density. The effectiveness of the clothing depends, as mentioned in 2.8, upon: (a) the type of wire and its gauge, (b) the angle formed at the knee, (c) the height of the knee and the point position, (d) the point density, and (e) the sharpness of the wire points. A steel wire of round cross-section is very popular; however wires of elliptical or double convex cross-sections are also used.

The whole wire is tempered, whereas the points are given special hardening treatment.

The gauge of the wire is chosen as per the point density required. A finer gauge and higher number of points/mm are used for longer and finer cottons. This is because the number of fibres for the same lap weight is greater for finer fibres. The various particulars of count of wire, corresponding gauges used and point density are given in Table 2.2

2.8.3   CARDING ACTION AND CARDING ANGLE[1]

When the fibres are carded between cylinder and flat, many of them get hooked around the wires of both surfaces. This causes drag, which is experienced by the wire points and leads to their tilting back a little. With PQRS (Figure 2.33) as the

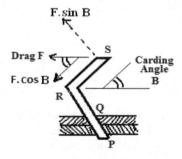

FIGURE 2.33   Forces on cylinder wire[1,4]: The wires carryout the carding action. This creates a force on the tip of the wires. The bending of wire under this force depends on both the height of the teeth and the carding angle.

cylinder wire, the component of dragging force (component—F. sin B perpendicular to SR), while displacing point S results in slight lifting. The raising of point S in relation to the base of the foundation actually brings it closer to the flat wire points. During working, as the dragging force is experienced by all the wire points on the cylinder surface, this *point-rise* is very likely to lead to friction between the cylinder and flat wires. Therefore, apart from wear and tear, this touching would lead to occasional sparking and cause fires.

However, the wire is flexible and so is the foundation, which is resilient, as well. Consequently, the component F. cos B (acting along and in the direction SR) has a tendency to lower point S. Ultimately, it all depends on the magnitude of both these forces.

Whether point S would rise or lower (as can be seen from Figure 2.33), is solely governed by angle B. As it is, the loading on the cylinder can displace the wires significantly. However, as they are very closely spaced, they do not have much allowance to bend through more angle. The cotton fibres also do not have unlimited strength and are likely to break before displacing the wires to an appreciable extent.

When it is necessary to get more opening power, smaller carding angle (angle B) is chosen. This increases the power of penetration of the wire points into entangled cotton clusters. In some of the flexible wires, the foundation is brought very close to the knee or right up to it (non-strip flexible clothing–buried knee) to give more support to the wires.

2.8.4 COUNT OF CARD CLOTHING[1]

1. Count the number of crowns per inch across the width of the fillet (e.g., 4).
2. A plain set will have a vertical repeat over two rows, whereas a rib set has a repeat over three rows. The repeat is called a *nogg*. Calculate the crowns per nogg.
3. Count the number of noggs or repeats in one inch along the length of the fillet (e.g., 25).

Thus, (a) × (b) × (c) = 4 × 3 × 25 = 300 crowns/in^2 (with rib pattern)

As there are two points per crown, 300 × 2 = 600 wire points/in^2 (2.3)

Using (2.3), we have

$$\text{Count of Clothing} = \frac{\text{Points per sq. inch}}{5} = \frac{600}{5} = 120$$

It was customary to use the count of cylinder clothing ranging from 80^s to 120^s depending on the class of work. For coarse cottons, a lower count was used. The clothing on doffer and flats, especially on doffer, were usually a ten count finer than that on cylinder. It thus provided more point density on doffer. On a card with flexible clothing, this was purposely so chosen as to help easy transfer of fibres from the

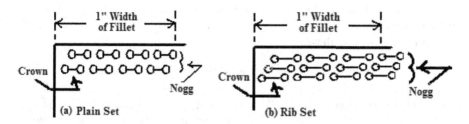

FIGURE 2.34 Pattern of crowns in flexible wire[1,4]: Depending on the basic weave arrangement, the crowns and therefore the noggs are formed. The wires are pierced accordingly. The procedure is continued along the whole length of foundation strip. (a) Plain set. (b) Rib set.

cylinder onto the doffer. It may be mentioned here that the plain weave pattern (Figure 2.34a and b) gives the highest point density amongst all the patterns. As against this, it may be noted that in metallic clothing, point density on cylinder is much higher than that on doffer. This is done to improve carding action. Further, the fibre-holding power of metallic clothing is very low, so the transfer of fibres from the cylinder onto the doffer is comparatively quite easier.

2.8.5 METALLIC WIRE CLOTHING[2,5]

The real success in the development of increasing card production was owing to the introduction of metallic wires. In the manufacturing process (Figure 2.35), a wire with a round section is flattened and rolled into a long strip comprising (a) a rib to constitute the base of the wire, and (b) a thin web forming a portion for teeth. This thin portion is stamped or punched so as to leave a thin serrated strip called *saw tooth wire*. It is also frequently called *rigid* wire—an apt designation. The tooth thus formed is very solid, strong and sturdy.

The teeth are usually subjected to a special hardening treatment. It involves controlled heating followed by cooling. This enables the retention of sharp wire points for a longer duration. The base or the rib portion, however, is specially tempered so as to partially retain flexibility. This is essential because; the wires, when mounted on cylinder or doffer, have to conform to their peripheries. Unlike licker-in wire, the

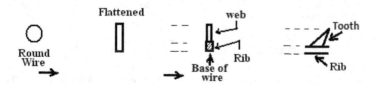

FIGURE 2.35 Transformation of wire[2,4]: Initially a round wire is used to make metallic saw tooth wire. Subsequently, the wire is transformed in steps to make it a flat wire of required thickness. Punching completes the transformation into saw tooth wire.

wires on the cylinder and doffer are simply put on their bare* surface. However, like metallic wires on the licker-in, both the teeth and the base of these wires form an integral part of saw tooth wire.

2.8.5.1 Metallic Wires on Cylinder and Doffer

The metallic wires on both the cylinder and the doffer are finer than the licker-in wires. Those on cylinder, especially, have much finer wire points to give quite a high wire-point density for maximum carding action. The basic difference between the cylinder and doffer wires is in their size and shape. The wires on the cylinder are shorter. Additionally, the point density on the cylinder, unlike the case of flexible wires, is much higher than that on the doffer. As for the thickness and height of the rib (the base), the cylinder wire is thinner and shorter. Angles A and B (Figure 2.36) are related to the carding action. As can be seen, they are complementary angles.

In short, these angles for the cylinder (angle A—of leading edge from the vertical, or carding angle B—from the horizontal) are different than those for the doffer. For the cylinder, angle B is greater than that for the doffer (and vice versa, angle A is less than that for the doffer). All these differences are due to the type of work that either cylinder or doffer is expected to do.

The mounting procedure for the metallic wires on both cylinder and doffer is similar; however the pressure on the wires during mounting differs. The pressure is required to be more when mounting cylinder wires.

Whereas there is a comparatively higher load on cylinder wires due carding action, the wires on the doffer merely collect and condense the fibres. The smaller size of the cylinder wires gives the required degree of point density and helps in giving a better carding action.

Thus, the higher wire point density helps in improving better fibre-to-fibre separation. The base of cylinder wire (thickness of rib) provides a firm seating. However,

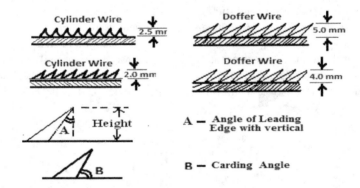

FIGURE 2.36 Cylinder and doffer wires[1,4]: The functions of cylinder wires and doffer wires are different. The former has to play a most important role in providing carding action, whereas the latter has to collect and retain the web received from cylinder.

---

* In the case of converted card from flexible to metallic, additional foundation wire was required to be put on the bare cylinder.

as far as the angle of the leading edge of the tooth is concerned, a wider angle (from vertical) while providing more intensive carding action also has more powerful hold on the fibres. This makes their transfer on to the doffer a little more difficult.

2.8.5.2 Types of Wires[5]

There are different types of wires available for both cylinder and doffer. These have sound and well-proven basic wire designs suitable for most of the carding conditions. The choice and selection of the wire, therefore, depends on the manufacturing conditions prevailing in the mill. Some of the factors governing the choice are as follows:

- Type of the material to be processed
- The staple length, the micronaire or the denier of the fibres
- Type of the card—whether conventional or high speed
- The cylinder speed, the production rate and the end use (whether carded or combed)

2.8.5.2.1 Cylinder Wire

The cylinder wires have very short teeth with heights ranging from 1.5 mm to 2.0 mm. The shape of wires is modified so as to avoid clogging of vegetable matter or seed coat fragments. For improving quality of carding and to enable processing of man-made fibres, a typical cylinder wire has a height up to 2.5 mm. For coarser man-made fibres (higher dtex), a wire with still greater height may be used. The leading angle of tooth (from the vertical) of cylinder wire, on a conventional card, is comparatively low ($12°–15°$), whereas the one for the high production card has slightly wider angle ($15°–25°$).

In the market, special steel alloy wires are available that can work at very high production rate, up to 80 kg/h, and yet offer quite an extended life. A typical wire with 0.4 mm base width and $35°$ angle of leading edge can offer as high as 1080 points/in^2. To suit different purposes, there are wires available with angles varying from $15°$ to $30°$ and offering point density from 400 to 1000 points/in^2.

Lakshmi has come out with Cutty-Sharp alloy steel for improved strength and wear resistance. A modern card wire uses refined, high-carbon steel, blended with alloying elements. This improves its resistance to wear. Along with this, a unique load distribution reduces wear and tear on the teeth, thus enabling longer life. There is a variety of products available with state-of-the-art technology to improve performance. In all such cases, the point density can be varied from 760 to 950 with an angle varying from $15°$ to $25°$.

The combination of carbon steel and other alloy elements makes the wire suitable for high production. They have a substantially longer life and it is possible to process as much as 360,000 kg of material before a new wire becomes necessary. In such types of wire, the height is around 2.8 mm and can be varied. Similarly, it is possible to vary the rib width from 0.65 mm to 0.8 mm and the leading edge angle from $14°$ to $20°$ ($66°$ to $76°$ from horizontal). This offers a point density from 500 to 760 points/in^2. There are other standard wires available for semi-high-production cards (converted metallic cards). With similar wire particulars, very small changes are made in the leading edge angle (10°, 15° and 20°) and points per square inch (450 to 760).

With the higher rates of production associated with a modern card, it becomes necessary to change the design of wires so as to strengthen the wire tip. This requires improved metallurgy and state-of-the-art technology to build consistent higher hardness and high wear resistance. It is possible to use a micro-alloy to substantially improve the life of cylinder wire. When newly mounted, these wires can process up to 1000 tons of material during their lifetime. Owing to their micro-structure, these wires have a very fine grain structure. Also, they are built with special tooth geometry and are given advanced heat treatment for improved hardness of the tooth. The wire is available in two different heights (2.0 mm and 2.5 mm) and four different rib widths (0.4 mm, 0.5 mm, 0.6 mm and 0.7 mm), with leading angle varying from 10° to 15° and points ranging from 600 to 1100 /in².

2.8.5.2.2   Doffer Wire

The purpose of doffer wire is totally different from that of the cylinder. There is no stress on the wires, as the important carding action has already been taken care of by the cylinder. In fact, the teeth on the doffer have only to collect the fibrous carded material from cylinder.

The appropriate changes in the doffer wires are accordingly made (Figure 2.37). The wire has to hold the transferred material firmly and quickly take it away from the influence of the fast moving cylinder wire surface. This requires a change in tooth geometry—the height, the angle of leading edge, the tooth spacing, etc. The taller doffer wires provide a longer reach to receive the fibres released from the cylinder. As against the 12° to 15° leading angle (from the vertical) of cylinder wires, the doffer wires are inclined to 20°–25°. Whereas with smaller angle, the cylinder wires are able to easily release the fibres, a comparatively larger angle of doffer wire helps it to catch, hold and retain the fibres, once transferred. As regards the point density, unlike cylinder wires, the doffer does not have to perform the carding action. Hence, it is not necessary to have a much higher point density on the doffer. Typical doffer wires have heights varying from 4.0 mm–5.0 mm, rib thickness from 0.8 to 0.9 mm, angle of leading edge from 20° to 25° and point density from 340–450 points/in².

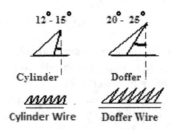

FIGURE 2.37   Comparison[2,4]: Cylinder and doffer wires differ in their height, angle of inclination and linear point density.

The *aero-doffer* wire for smooth transfer of fibres from the cylinder onto the doffer has also been developed. Here the tip is designed to assist easy transfer. The new design helps in avoiding turbulence of high velocity of air at the cylinder–doffer junction. Another special wire has a height of about 4.0 mm, leading angle of around 35° and nearly 320 points/in². Yet another wire in the same series maintains the height and angle but offers 276 points/in².

Sharp Mate and Exotic are the typical wires from Lakshmi. All such wire manufacturers offer a variety of wires with varying wire particulars. This is basically to suit the type of material processed. Thus, the doffer wires varying in height (4–5 mm), rib width (0.9 and 1.0 mm) and tooth angle (25°–30°) are available. It is also possible to vary the point density (300 to 390) depending on the class of work. The values quoted above are only the guidelines, and depending on the manufacturers of card wires, the actual values vary.

2.8.5.3 Wire Geometry and Material Processed[5]

As mentioned earlier, the important parameters of card wires are as shown in Figure 2.38. In carding, the wire geometry plays an important role. The wires on licker-in, cylinder and doffer are chosen depending on the staple length or trash content in the material and whether it is cotton or man-made fibres.

1. Pitch of the tooth
2. Height of the tooth
3. Rib width or base width
4. Tip width
5. Carding angle or front Angle
6. Back Angle

1. **Pitch:** The pitch of the tooth is inversely proportional to the linear density of the teeth. This actually means the space between two adjacent teeth, and this determines the tendency for material to clog between the teeth. The pitch also decides the point density (points per square inch) which, in turn, decides the intensity of the wire-tooth action. Point density is also a factor to be considered when processing coarser or finer fibres. For the same lap weight, the number of fibres to be treated is higher with finer fibres. In this

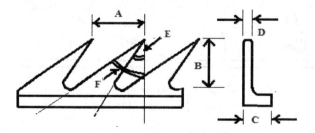

FIGURE 2.38 Wire geometry[1,4]: Factors like pitch, height, base width and carding angle all govern the power of the action of teeth in carrying out the work assigned to them.

case, the density of the wire points must be higher so as to maintain the same level of carding action.

2. **Height (or Depth):** The height of the wire decides the reach and fibre carrying capacity of teeth. The licker-in wire has to plunge into cotton sheet, to open the tufts and separate the trash. The height of the wire is, therefore, greater. As against this, a much finer carding action is carried-out by cylinder wire tips (land). However, the fibres are not supposed to enter too deeply into the clothing. The height, in this case, is the least among the three types of wires. As mentioned earlier, the reach is more important in the case of doffer wires and their height is more than that of the cylinder.

3. **Rib/Base Width:** The rib gives strength to the wire. At the same time, it has also to conform to the periphery of the cylinder or doffer. In the case of the licker-in, however, the wire has to be fitted into the grooves. Accordingly, the appropriate rib width is chosen. As the wires are mounted on bare cylinder or doffer, their sides touch each other. In this case, therefore, the rib width decides the space that each wire occupies when it is placed touching its neighbouring wire. In a way, it governs point density on the surface. For enhanced carding action, it is necessary to have much higher point population on the cylinder. The rib width of cylinder wire, therefore, is again the least amongst the three types of wires.

4. **Carding Angle:** As mentioned earlier, it is the angle of the leading edge (Figure 2.39) of the tooth with the vertical, this angle was measured with reference to the horizontal. While deciding the penetrating power of the tooth, it also controls its retaining power. For example, the wider the wire angle (from the vertical), the more powerful its teeth penetrate into the small cotton tufts. With the licker-in, wider angled wires are used when processing short staple, trashy cottons. As the wider angle also gives more retaining power to the teeth, the fibres are held more firmly by the teeth. This may not be advisable while processing long staple cottons, which normally have much less trash content. Even while processing man-made fibres, which have no trash content, the same is true.

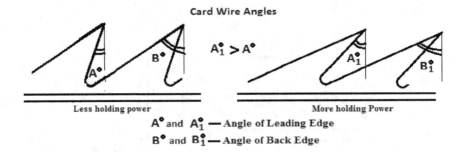

Card Wire Angles

$A_1^\circ > A^\circ$

Less holding power   More holding Power

A° and A_1° — Angle of Leading Edge

B° and B_1° — Angle of Back Edge

FIGURE 2.39   Angle of card wire[1,4]: The smaller angle of the leading edge imparts less hold on the fibres. All wires for the cylinder are made using this principle. As against this, doffer wires require more retention of the fibres that are transferred from the cylinder.

The carding angle of wires in this case can be perpendicular or even slightly negative. Between cylinder and doffer, as stated earlier, the wires on the latter have a comparatively wider angle simply to increase their retaining power. With the cylinder, however, a comparatively narrower angle ensures that the fibres are quickly released onto the doffer at the cylinder–doffer junction point. It may also be noted that with too narrow angle of cylinder wires, the control over the fibres is lost and this may lead to increase in the flat strip, especially at higher cylinder speeds owing to centrifugal force experienced by the fibres.

5. **Angle of the Back Edge:** This decides the fibre loading tendency and it is also responsible for the strength of the teeth. The lower angle reduces fibre loading, whereas, with the higher back angle, penetration of the teeth into fibre tufts is assisted. The best angle, therefore, is the one which balances both these opposite characteristics.

2.8.5.4   Types of Foundation Wire[2]

When a card, originally clothed with flexible wire, is to be converted to metallic wire clothing, an additional wire at the base called the *foundation wire* is required to be mounted first on the bare cylinder.

This is necessary because of the difference in the height of the two types of wires. The height of the cylinder flexible wire is 9 mm (0.375 in) from the base of the foundation, whereas the height of metallic wire on the cylinder is only 3 mm (0.125 in). Therefore, the height of the foundation wire is chosen to compensate for this difference. The castellated design (Figure 2.40) of this wire permits the requisite expansion over the diametric measurements when seated on the bare cylinder surface. The punched wire also reduces the weight on the cylinder. Though a steel based wire is ideal for precision work, it unnecessarily adds an extra load on the cylinder (Table 2.3). The heavy strain is thus put on the walls of the cylinder, its shafting and bearings.

This extra load coupled with centrifugal force due to cylinder speed leads to its deformation, especially when the old flexible wire card, after being converted to metallic clothing, is run at higher cylinder speed. An aluminium base wire in place of steel base wire was then found very suitable owing to its lighter weight. Aluminium, however, has a high coefficient of expansion than steel and care had to be taken while grinding.

(a) Metallic Wire Base   (b) Plastic Base Wire

FIGURE 2.40   Base wire for cylinder and doffer[2]: The important requirement of the base wire is that it should conform well over the curved surfaces. It should be strong and should not distort or stretch under the mounting tension. A metallic wire is definitely strong but it adds to the total weight of the cylinder. A plastic wire is lighter but there is danger of its distortion and it can easily succumb to the mounting pressure. (a) Metallic base wire and (b) plastic base wire.

TABLE 2.3

Type of Base Wire and Increase in Weight

No.	Type of Wire	Increase in Weight (kg)
1.	Steel Base Wire	175
2.	Aluminium Base Wire	110
3.	Plastic Base Wire	42

The introduction of plastic base wire substantially reduced the problem of weight. The mounting technique for plastic base wire is the same as that used for other types. Another advantage of using plastic wire is that the wire had no expansion on heating at the time of grinding. Hence, there was no risk of distortion. However, it was necessary to wind such wires with a certain minimum tension. The normal tension during card wire mounting ranges between 18 and 20 kg. The plastic wire, being comparatively weak, required that a lower tension be used during mounting. However, if due care was not taken, this was likely to allow small pockets of air between the bottom of the wire and bare cylinder. Thus, when the metallic wire was subsequently mounted on this bed of foundation wire, it led to uneven mounting of a new card wire on the cylinder or doffer. It is because of this that plastic base wire did not gain popularity. It may be noted that no base wire is required when the card is originally manufactured as metallic wire card.

2.8.5.5 Flexible versus Metallic Card Clothing[2]

The very nature of the work involved necessitates very strong and sturdy metallic wires on the licker-in. Even then, when the metallic wire on cylinder and doffer was first introduced, many cotton technologists preferred flexible wire for fine and superfine mixings, believing that the required point density could only be met with such clothing. But as the technology of higher card production developed, there were improvements in metallic wire manufacturing as the use of metallic wire on both cylinder and doffer became increasingly essential. Though the basic functions of card have never changed, very high production rates were possible only with metallic wires. The merits and de-merits of them are discussed below:

1. The metallic wire clothing has saw-tooth wires and the teeth formed are integral part of the base or foundation rib. Hence the wire is quite solid and sturdy in structure. Both the teeth and the foundation are rigid; so, there is no danger of displacement of wire points during carding. The flexible wire clothing, on the other hand, has wires separately inserted into the foundation material. The wire points (teeth) and the foundation are both flexible. The characteristic feature of the clothing is the bent wires forming "knees". The flexible wires get displaced owing to the drag that they experience during carding. The resulting displacement, though limited, involves the risk of point rise.
2. Because of the firm structure of the teeth of metallic wires, any carding angle can be chosen. However, the risk of point rise with flexible wires limits the choice.

3. A typical characteristic of metallic wire obviates the necessity of frequent stripping. The fibre-holding power of these wires is very low, and as such, there is no accumulation of fibres in the clothing. All that is necessary is to apply brushing occasionally to remove broken seed fragments and other impurities embedded in between the wire teeth. It is therefore claimed that about 3% of the fibres, which otherwise would go as stripping waste, can be saved. In flexible wire, however, the bent knee leads to substantial amount of fibre accumulation. Though this also includes trash and other impurities, the wire points become clogged after a comparatively shorter interval. This loading of fibres into clothing seriously impairs the continued carding action of the wires. Hence, a frequent stripping action is required to clean the clothing. Apart from the time lost in this operation, there is a significant variation in the carding action. The rate of absorption of the fibres in the clothing, immediately after the stripping, is much higher than that in the later stage. Consequently, during this period, the hank of the sliver delivered also fluctuates.

4. The main carding action, involving individualization of fibres, depends on the action of wires in disentangling the fibre clusters. It was claimed that not only the requisite point density could be obtained in flexible-wire clothing, but also that the unique action of flexible wire points in carrying out carding could not be excelled by saw-tooth points. As far as point density is concerned, the latest types of metallic wires do provide very high number of points per unit area.

5. In a conventional flexible-wire card, no separate foundation wire was required for mounting the wire fillet; however, after converting to metallic wire, foundation wire became essential to compensate for the difference in their heights. As mentioned earlier, foundation wire is not necessary for the card originally manufactured as a metallic wire card.

6. Saw-tooth construction being quite sturdy, the metallic wire points lose their sharpness very slowly. It is normal to grind these wire points only after 8–11 months, depending on the class of the work carried out.

   In the case of flexible wires, however, it is the point which is mainly responsible for carding action. These wires being comparatively softer often get dull after a short interval of 2–4 weeks. This involves grinding, which leads to a considerable loss of productive time. One interesting point to note is that when newly mounted, both the flexible and metallic wires present a fine point for carrying-out a carding action (Figure 2.41a and b). However, after grinding, the metallic wire no longer presents this fine point. In fact, its top surface becomes flatter after each grinding.

   The flexible wire, however, after each grinding continues to present a fine point, thus offering the same sharpness for repeated carding action. But it has been proved that the flatter top of metallic wires, after each grinding, is still very effective in carrying out the carding action.

7. The life of flexible wire, owing to repeated grinding, is limited. This is because only a part of the length above the knee is available for carding action. After each grinding, this length is continually shortened, thus

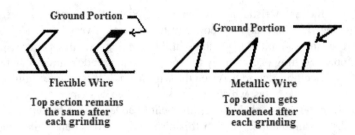

FIGURE 2.41 Effect of grinding on flexible and metallic wire[2,4]: The flexible wire is round or slightly ovular in cross section. When the points are ground, there is reduction only in height; the cross-sectional shape remains the same. With each grinding, the sharp point at the top of the new metallic wire is broadened. This point turns into small rectangle.

limiting its usage over a longer time. The wearing of metallic wires owing to carding action is quite slow, and hence, its life is appreciably longer.

8. The mending of flexible wire is done by replacing the damaged portion, but the operation requires lot of skill. With metallic wire, the damaged portion is replaced by replacing the wire, which is spot welded. In this case, removal of the complete wire is not necessary. However, in both, the process of replacement is quite time consuming and requires great care.

9. Special types of flexible wires are introduced to save the time lost in their frequent stripping. They are known as *strip-less fillets*. These wires have the knee portion submerged just below the surface of the foundation. Further, the special steel wires which were introduced in flexible fillet manufacturing reduced the grinding frequency appreciably.

2.8.6 SPECIAL FLEXIBLE CLOTHING[1,2]

These wires are not flexible in the true sense and can only be called *semi-rigid*. In the conventional flexible wire fillet, new varieties were introduced mainly to eliminate the need for frequent stripping. Basically the idea in such developments was to reduce the fibre loading tendency. In addition, the wire points were given special hardening treatment to increase their life.

The main disadvantage of flexible clothing was that it used to absorb fibres. Once the fibres were embedded into the clothing, the bent knee made it difficult for the fibres to rise again to the surface. The advantage of semi-rigid clothings was that there was no knee (or it was submerged in the foundation). The foundation was also changed accordingly to increase its rigidity. The semi-rigid wire, apart from a hardening treatment, was given a special finish. The wires were thus sharper and had finer points.

It was claimed that with this special clothing, there was substantial reduction in neps. The side grinding on either side of the wire points enabled improved carding action. The hardened points also retained their sharpness for a longer duration, thus extending the grinding cycle. As a result of all this, there was remarkable improvement in the web quality.

2.9 CYLINDER BENDS[1,2]

On each side of the cylinder, strong cast-iron arches known as cylinder bends (Figure 2.42) are securely fastened to the framing. Each bend carries six flat setting brackets and another set of brackets holding stripping and grinding rollers. On the lower side, the bend has a pronounced rib which supports the adjusting set screws. The cylinder is then closed from the sides at the flanges by metal sheets. This helps in shielding the fibres between the cylinder and flats from the generated stray air currents which while trying to go out from the sides are likely to disturb the carding action.

2.9.1 FLEXIBLE BENDS[1,2]

Flexible bends directly support the bearing ends of the flats on either side and allow the flats to slide over them. Each bend is approximately 150 cm (60 in) long and about 2.5 cm (1 in) wide. A taper is seen from its central position to either side.

The bends are not flexible in the usual sense because they can't be extended along the length. However, their curvature can be altered within reasonable limits. This is to make them conform to the periphery of the clothed cylinder. When the flexible wire cards were converted to metallic wire cards, the change in the cylinder curvature owing to reduced height of the metallic wire demanded more bending of these flexible bends beyond their normal limit. Obviously, this was not possible. This was the main reason for the use of foundation wire when the cards were converted from flexible wire to metallic wire.

It may be noted that, even with metallic cards, each grinding operation reduces the height of metallic wire. Here, too, the curvature of the cylinder is changed. This necessitates re-conforming flexible bends to the new reduced diameter (and the curvature) of the cylinder. It ensures accurate and uniform setting of the flats with the cylinder at all the working surfaces. But the curving of the flexible bends required in this case is perfectly within its normal elastic limit of the bends. Each flexible bend is supported on five adjusting brackets (Figure 2.43b).

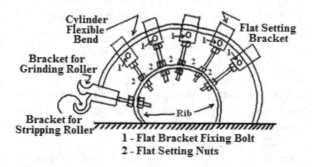

1 - Flat Bracket Fixing Bolt
2 - Flat Setting Nuts

FIGURE 2.42 Cylinder bends with flat setting points[1,2]: The flexible bends are positioned over the two main arcs placed on either side of the cylinder. The bends are flexible in the sense that their curvature can be slightly changed to conform to the circular cylinder surface. This is essential while carrying out the flat setting. 1. Flat bracket fixing bolt. 2. Flat setting nuts.

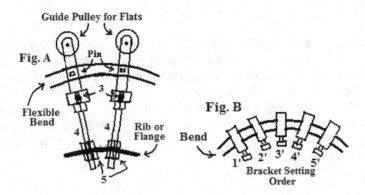

FIGURE 2.43   (A and B) Flexible bends[1,2]: The bends are supported by the setting brackets. There is a standard procedure to set the flexible bends around the cylinder so that there is no strain when their curvature is changed.

The strong pins held by the flexible bend projects through the horizontal slots in the brackets.

A bolt-3 (Figure 2.43A) passing through a vertical slot of the bracket is secured to the cylinder bend. The set screws-4 (Figure 2.43A) extend from bottom of the bracket and pass through a rib or a flange. They are securely held by the lock nuts (5). While carrying out the setting of flats with the cylinder, first the bolt (3) is loosened and then the required movement of the flat setting bracket (and therefore flats) with the cylinder is effected by using lock nuts 4 and 5 suitably placed on either side of the rib.

Whereas the movement of the central bracket 3' (Figure 2.43B), results in the direct change in the distance between the cylinder and flats, the movement of the brackets on either side partly changes this distance and partly alters the curvature of the bends. The movement of the brackets, therefore, has to be carried out in a certain manner. The setting may be initiated from one of the ends and then serially carried out on each subsequent bracket (e.g., 1' → 2' → 3' → 4' → 5') or it may be started with the central bracket (3') and then the two brackets adjacent to it (2' and 4'), and then finally the extreme brackets(1' and 5'). This ensures a uniform change in the curvature of the flexible bend over its entire length and also avoids any fatigue owing to change in the curvature of the bend.

2.9.2   Types of Flexible Bends

The rigid bend A (Figure 2.44) in Howard & Bullough is mounted on the main arch F. The upper surface of the rigid bend is bevelled so that it tapers-off towards the cylinder. The rigid bend is held in position by five bolts and on each of these is positioned an *index nut B.* (Figure 2.44).

The index nut is placed on the outer side of the rigid bend A; whereas a threaded screw passes through it and carries lock-nut C from inside. The flexible bends support the flats and rest on the bevelled surface of the rigid bend A. The weight of the flats acts on the flexible bend which, in turn, is held against

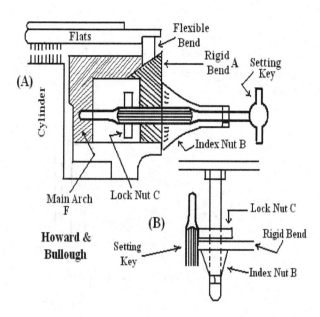

FIGURE 2.44   Howard & Bullough.

the projecting main arch F. Therefore, when the rigid bend is made to slide away from the cylinder, the flexible bend falls down, thus closing the distance between the cylinder and flats. The setting is carried out by initially loosening locknut C and then bringing the required movement of the rigid bend through index nut B. As this setting must be carried out at five points, where the flat setting brackets are provided, the number of index marks through which index nut B is turned at one place can be used as a guide line for setting the other brackets. For this, the dial on index nut B is graduated so that each division indicates 0.0254 mm (0.001 in) rise or fall of flexible bend. It is important to fasten the locking nut after the setting is completed.

In Tweedle & Smalley's attachment, the flexible bend A is supported on the regulating screw B (Figure 2.45). The screw carries an index nut C. The fixed bend D is accordingly grooved to receive the flexible bend at the top and the regulating screw at the bottom. Another locking screw E passes through the flexible bend and is held at its lower end by forked-nut bracket F. To carry out any adjustment in the setting, the locking screw E is first loosened. The index nut C is then turned so as to raise or lower the regulating screw B.

The flexible bend A thus moves accordingly and the required setting between cylinder and flat is obtained. As in the case of Howard & Bullough, the minimum distance through which the flats can be moved in Tweedle & Smalley is also 0.0254 mm (0.001 in).

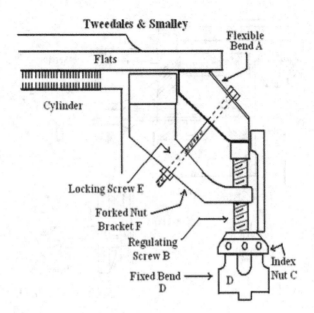

FIGURE 2.45   Tweedle & Smalley.

2.10   FLATS[1,3]

Each flat is primarily a cast iron bar of T shape cross section (Figure 2.46A and B). The upright stem comprises a strengthening rib and the horizontal cross bar, the latter constituting the working surface of flats. The flat clothing is mounted on this working surface. The upright strengthening rib does not extend to the extreme ends of the cross bar; some room is provided to allow the joining of the flats with chain. There are, in general, 105–110 flats, each about 35 mm broad (1⅜ in). However, on working surfaces over the cylinder, there are only 45–50 flats at any time. This constitutes about 40–42% of the total flats in working position over the cylinder at any time.

There are two distinct portions on any flat: (a) a working surface clothed with the wires, and (b) the bearing surface specially shaped to have a concave face. The purpose of this concave face of the bearing end (Figure 2.47A) is to provide a firm

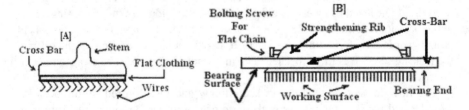

FIGURE 2.46   (A and B) Flats[1,2]: The flats have a typical cross bar with a strengthening rib (T-shape stem) and carry the wire clothing called *tops*. The extended surface beyond the wire clothing is a bearing surface which is made to slide smoothly over the flexible bends.

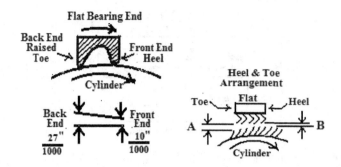

FIGURE 2.47   Heel and toe[1,4]: This is a special designing arrangement of the bearing surface of the flats. It automatically gives the gradient from entry point to the exit point of for each flat. The actual setting between the flats and the cylinder is done at the exit point (front end). (A) Heel and toe arrangement. (B) Gradually intensified carding.

support (seating) to the flats at two distinct points. Thus, the flats during their journey over the flexible bends are able to move with a firm supporting balance at their bearing ends.

It is obvious that neither a straight face nor a matching curved face corresponding to the curvature of the flexible bend could really have served this purpose. With the former, the flats would never have been correctly balanced on a curved surface of flexible bend. In the case of the latter, the balance would have been lost when the curvature of the flexible bend was altered after every subsequent setting between cylinder and flat. This is because, after every grinding operation, the reduction in the cylinder diameter would necessitate resetting of the distance between the flats and cylinder. It may also be noted that the back end of the bearing surface of the flats is slightly raised.

This tilts the working surface of the flats in such a way that the fibres enter each flat-cylinder region through a wider gap, A (Figure 2.47B). As the fibres move through each flat, the distance between the flat wires and cylinder progressively narrows down to B (Figure 2.47B). Thus, if the back end of the flat (entry of fibres) is at 0.685 mm (27/1000 in), its front (exit of the fibres) end is set at 0.254 mm (10/1000 in). The arrangement of each flat thus offering a progressively narrowed distance from the entry to the exit is called a *heel and toe* arrangement. The heel is the leading portion and the toe is the trailing portion of each flat. This arrangement avoids any suddenness in the carding action.

The fibres carried by the cylinder at a relatively high speed are gradually brought through this space gradient from A to B (Figure 2.47B). As each flat is designed and set in the same manner, the severity in carding action is reduced at every entry point of the flats. Thus, the fibres pass through each cylinder–flat region and the gradual closing of the distance (from A to B) maintains the same effectiveness in fibre–tuft opening action. It may be noted that the usual setting (of 0.254 mm or 0.010 in) between the cylinder and the flats, however, is carried out at heel only. The distance at the toe (0.685 mm or 27/1000 in) is automatically obtained owing to the heel-and-toe arrangement.

In the conventional card, the cotton is transported by cylinder wires at a speed of 630 m/min (nearly 2100 ft/min) past 45–50 flats, which themselves move at a

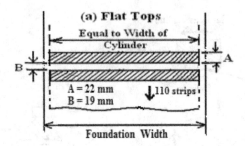

FIGURE 2.48   Flat clothing.

comparatively very slow speed of 7.5 cm/min (3 in/min). Each flat is clothed with a wire strip called *flat top*. These tops are mounted on the working face of the flats by special clips. With each top having a width of 22 mm (7/8 in), a total flat wire surface of approximately 100 cm is available for carding action at any time in one rotation of the cylinder. As the cylinder wires are inclined in the forward direction and the flat wires in the backward direction (to their normal motion), a point-to-point action is experienced by the fibres throughout the working area of flats. In fact, this is the real carding action and it continues between each flat and the cylinder over the entire working area between cylinder and flats.

Thus, there is a constant exchange of fibres from the cylinder onto the flats and back onto the cylinder. The fibres on the cylinder tend to rise slightly to the surface between gaps of each flat and thus get repeatedly and thoroughly carded. Factors such as air currents, centrifugal force and the elastic nature of fibres all help in bringing about this action.

2.10.1   FLAT CLOTHING

The production of flat tops is carried out in sets, each containing as many strips of the tops as there are total number of flats on a card. The machine inserts staple wire rows across almost the full width of the foundation, leaving only a small margin at both sides (Figure 2.48). After every 22 mm width, the material advances by 19 mm before starting for the next top. The whole set of the tops is then uniformly ground and eventually cut into individual tops-strip. The foundation for the tops is made in the same manner as described earlier. The tops can be flexible or semi-rigid (Figure 2.49) with the knee buried.

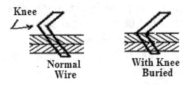

FIGURE 2.49   Flexible wire foundation.

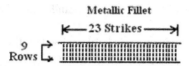

FIGURE 2.50   Metallic tops.

Each wire strip is a complete flat top, which later is mounted underneath the cross bars of each flat.

2.10.2   METALLIC WIRES ON FLATS[1]

The metallic wires are mounted on the flats in such a way that the teeth are arranged in a regular pattern or as a set, sometimes in 9 transverse rows. Some manufacturers use 11 transverse rows with 23 longitudinal rows or *strikes* (Figure 2.50) giving a tooth density of 32 points per square cm (216 points/in^2). The overall height of the teeth is 5.7 mm (0.225 in) and the carding angle is 20° to 22° (from vertical). This carding angle is appreciably different than that for flexible clothing on flats (Figure 2.51a and b).

2.10.3   FLAT TOPS FOR HIGHER PRODUCTION[6]

Several varieties of flat tops are now available. These tops are found suitable when there is metallic wire on cylinder and doffer. Thus, ICCO and Cresta tops are very popular at semi-high and high production rates. At still higher rates of production, Rappo HS$_2$ tops have replaced Cresta Diamond hard tops. These are similar to Eureka tops and are used for synthetic fibres.

2.10.4   TYPICAL FLAT TOPS[1]

These tops are specially designed to give better resistance to the damaging influence of trash particles. Depending on the trash content and whether the processed cotton is short, medium or long staple, special types of flat wires are designed with varying tooth densities (points per square inch). Even the choice

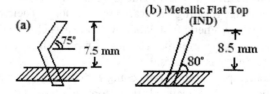

FIGURE 2.51   (a) Flat tops and (b) wire angle.

is available on the basis of production rates and type of fibre processed (cotton, viscose or polyester).

The teeth are side ground and extra hardened so that the sharpness is retained for a longer time without needing grinding. However, limited and controlled grinding is possible. The main advantage of some of the tops is that they ensure a minimum flat strip, which is one of the most important requirements when processing man-made fibres (no trash).

2.10.5 Cresta Tops

In this, the wires are semi-rigid type with a knee arrangement. The points are shorter and harder and require very light grinding after an interval of one year. It is capable of giving better carding results and takes out only slightly more waste than "strip-less" Eureka tops.

2.10.6 Eureka Tops

These tops have extra-hardened points with no knee. The foundation is thick and teeth are bevel-cut to give chisel-type sharp points. As tops extract about 50% less flat strip, they are suitable for processing man-made fibres. However, their power to hold the trash particles is equally striking. So they may also be very useful in extracting trash for medium or lower grade varieties. They are known as strip-less tops, owing to their nature in extracting much less flat strip. Because of the specially hardened teeth, they are not required to be ground for 2–3 years.

2.10.7 Some Other Flat Tops[6]

Lakshmi has come out with different types of tops with varying points per square Inch. They have the same height (8.0 mm). Some of these are: Supra tops (350–460 points/in^2), Prima tops (320–520 points/in^2), Eco tops (300–450 points/in^2) and Picco Diamond tops (330 points/in^2).

Graf has different types of tops: Resist-O tops (410–550 points/in^2), Prima tops (400–520 points/in^2) and Supra tops (350 points/in^2). Inline-X tops (350–500 points/in^2), Diamant/Picco Diamant (240–330 points/in^2) and Fully Metallic tops (310–420 points/in^2).

Lakshmi-Graf has some special tops which differ from ordinary tops. Their Rappo tops are set in two sections—widely spaced and closely spaced wires. This is expected to extract trash and give fine opening of the fibres (better carding action). With them, the grinding intervals are very long. Their Picco -Diamant tops have a reinforced synthetic foundation which resists chemical wear and does not absorb moisture, thus preventing rusting of wires. These types of wires are specially meant for man-made fibres.

ICC has developed special tops for matching with their cylinder wires, especially for high production rates. As claimed, these tops are expected to achieve maximum carding efficiency. Their Endura top range has a special setting pattern and

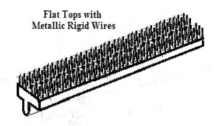

Flat Tops with
Metallic Rigid Wires

FIGURE 2.52   Split flat tops[2,4]: The idea of a split top was basically introduced to refresh carding action. Thus, with a single split, the second segment gives renewed fresh carding action.

distribution of wire points to maximize the carding power. The needle-sharp points have lengthened the grinding frequency, and at the same time, reduced neps and yarn faults (angle 15° to 20° and 350 to 530 points/in²). From the other varieties of ICC tops, their Xltops and Iccoflex tops are manufactured with the latest *back-off* grinding technology (240 to 430 points/in²).

2.10.8   SPLIT TOPS

These can be either flexible or metallic. The tops are divided (Figure 2.52), keeping the central portion blank. It is also possible to vary both the fineness and point density of each section so that the wire section in the back (towards entry side of each flat) is coarser and less densely packed. It is claimed that this increases their carding power. The flat strip is also reduced by about 20%, and there is significant nep removal.

2.11   FLAT STRIPPING COMB[1,2]

As the flats move slowly away from the cylinder wire surface, they carry quite a significant portion of fibrous matter on their wire surfaces. Eventually, when the flats are again brought back from the licker-in side to carry their renewed carding operation, it is necessary that they be cleaned. When the flats move around a small carrier wheel which takes them away from cylinder, their wires are bent in a downward direction (Figure 2.53). This is the best position for stripping the flat wires.

The flat stripping comb, which is a finely toothed blade, is made to oscillate in close proximity to the back side of the flat wires. This brings about the required stripping of fibres from the flat wires. The eccentric mounted on the flat driving wheel shaft (Figure 2.54) gives the rocking motion to the comb. The up and down motion of rod B about the fulcrum F along with the connection of the loose brackets C and D, results in yet another secondary motion given to the blades.

Thus, the comb blade T turns in the direction which helps in stripping the fibres from the back side of the flat wire surface. The setting between the comb and the flat wires may vary from 0.43 mm (15/100 into 22/100 in). Care, however, should be taken to see that at no time does the comb blade touch the flat wires. This obviates

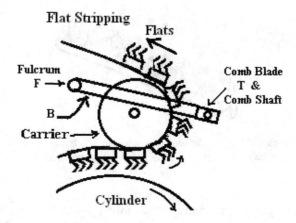

FIGURE 2.53 Flat comb[1,2]: Both the vibratory and the oscillatory motions are simultaneously imparted to the flat comb. This enables it to strip the flat wire surface more effectively.

any possibility of damage to flat wires. It may, however, be noted that this setting is not critical in the sense that it does not affect the quality of the material being processed. Even then, though, it merely helps in stripping the flat wires effectively; the flats, if not fully cleaned, will slowly buildup the gathered fibrous matter and, in the course of time, would become less effective in carrying—out carding action. The flat strip, consisting of embedded fibres, also carries trash and other vegetable matter.

In the past, after the strip was removed from the flats, it was allowed to fall and accumulate in the space between the cylinder and doffer over the covers. Subsequently, these strips were wound on a flannel-covered collecting roller in the form of a soft lap.

In either case, it is necessary remove the collected strip periodically. The comb, however, is not able to remove the fibres that are more deeply embedded between the flat wires. For this, an additional stripping and cleaning brush (Figure 2.55) with its bristles moderately penetrating the flat wires is provided. The brush is slowly rotated

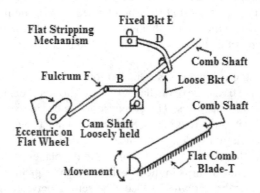

FIGURE 2.54 Flat comb[1,2]: Both the vibratory and the oscillatory motions are simultaneously imparted to the flat comb. This enables it to strip the flat wire surface more positively.

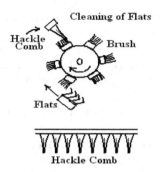

FIGURE 2.55   Cleaning of flats[1,4]: The brush bristles penetrate the flat wires and bodily lift the accumulated trashy fibrous matter. In turn, the hackle comb cleans the brush bristles.

with the help of a grooved pulley. The bristles, in turn, are cleaned by a coarse hackle comb. The cleaned flats are subsequently brought around to enter again from the feeding side (licker-in side) from where they recommence their journey over the cylinder wire surface for continued carding action. Thus, cleaned and fresh flats are continuously presented to the cylinder.

2.12   CYLINDER UNDERCASING[1,2]

The underside of the cylinder, from licker-in to doffer, is surrounded by an undercasing. The undercasing is divided into two semi-circular segments (Figure 2.56), AB and CD, which are made perfectly concentric to the cylinder curvature. The front side of the portion of this undercasing (on the doffer side), is supported at each side by brackets, the pin on these brackets passing through the slots at A. In the same manner, the back end of the front portion, AB, is supported at B. For the back

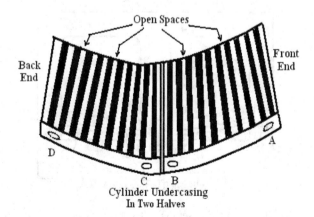

FIGURE 2.56   Cylinder undercasing[1,2]: This is mounted underneath the cylinder and extends almost from licker-in to doffer in the front. It streamlines the concentric air currents and supports the fibres on the cylinder.

portion, CD, similar arrangements are made to receive the pins from another set of the brackets at C and D respectively.

The setting of the cylinder undercasing on the licker-in side is related to the setting of the licker-in itself. For this, the licker-in undercasing is required to be set at 0.86 mm (34/1000″) near the cylinder side. Hence, the cylinder undercasing is accordingly set so that it is approximately at the same distance from the cylinder wire points (setting at D_1 in Figure 2.57). At C_1 and B_1, the undercasing is set a little wider (1.72 mm, 68/1000 in).

At the doffer side (A_1), the setting with cylinder wires is widest and is around 2.5–3.0 mm (0.125 in). Thus, there is a steady increase in the distance between the cylinder undercasing and the cylinder from D to A.

The two halves of the undercasing are set face to face at B_1 and C_1. Both the halves are provided with open spaces or slots running parallel along its width (Figure 2.57). The function of the cylinder undercasing is to control the movement of air currents around the cylinder. This is very important as, after the air currents continue their journey past the cylinder–doffer junction (which is the narrowest distance), they are likely to create turbulence, if not controlled. The undercasing also supports the fibres close to the cylinder wire surface and this prevents their falling down due to centrifugal force. The gradual closing of the setting from doffer side to licker-in side with a uniform gradient thus streamlines the surrounding air currents. The open spaces or slots in the two halves of the undercasing provide an outlet for any stray air currents to pass through, thus avoiding any turbulence in the space between cylinder and its undercasing. These open spaces also provide the exit path for fine trash particles and short fibres.

The working surface of the two halves of the undercasing (towards the cylinder) must be kept very smooth so as to avoid any disturbance to the flow of

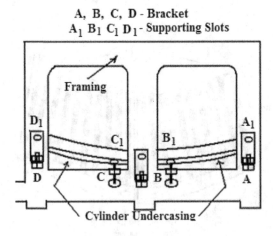

FIGURE 2.57 Cylinder undercasing setting[1,2]: The undercasing is in two halves, so provision is made to set it at four independent points. Even then a smooth gradient is required to be maintained from doffer side (point A_1) to licker-in side (point D_1). A, B, C, D: Bracket and A_1, B_1, C_1, D_1: Supporting slots.

air currents over them. Any roughness is likely to lead to air turbulence. The roughening also leads to frequent accumulation of fibres which are subsequently rolled into loose fibre balls. These fibre balls are delivered, almost in the same state, and appear in the final web formed at doffer. This typical defect is called *snowball formation.*

2.13   FRONT PLATE

This plate controls the air currents around the cylinder in the region where the flats start leaving the cylinder. Indirectly, it controls the percentage of fibres on the outgoing flats and therefore influences the flat strip percentage. Hence, the plate is often called the *percentage plate.* It is positioned at the point where the flats start moving away from the cylinder (Figure 2.58).

The setting of the plate is adjusted by means of set screws D and E, whereas bolt F fixes the plate to the concentric bend around the cylinder. While carrying out the setting of this plate, it is necessary to first loosen this bolt F. Set screws D and E are used to alter the setting at the back and front end of the plate respectively. Normally the back end (D) is set at 0.254 mm (10/1000 in) whereas; the front end is set at 0.762 mm (30/1000 in).

It is interesting to note that unlike any other setting in the card where the fibres are made to enter a wider distance and then passed through a gradient which closes the distance, with front plate it is the reverse. Thus, the fibres first pass through a comparatively narrow distance in between the cylinder wires at point D (Figure 2.59) and then this distance gets progressively wider till the point E. By changing the setting at D, the flat strip percentage can be varied within reasonable limits. Truly speaking, it is the gradient between D and E that is more important to decide the flat strip percentage, rather than the absolute value of the setting at D. Therefore, the position of E with respect to D really influences the flat strip percentage.

Further, any movement of either set screws—D or E—affects the setting at the other end. Consequently, when one edge of the plate is adjusted, it is necessary to

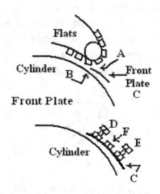

FIGURE 2.58   Positioning of front plate.

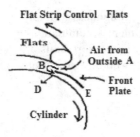

FIGURE 2.59   Control of flat strip percentage.

adjust the other end. This requires the use of both set screws to finally arrive at the correct distance of the plate ends from cylinder wires. The closer setting at D (Figure 2.59) allows the air currents to rush through its narrow space. However, the space at E is comparatively wide to reduce the air pressure and allow the partial vacuum to be created at E.

   This induces outside air currents to enter at A, turn around B and enter the narrow space between the back end of the front plate and cylinder. The path of this outside air is such that it takes some of the fibres from the flats back onto the cylinder surface. This action depends upon the strength of outside air currents rushing-in at A (Figure 2.60). The stronger these currents, the greater the proportion of fibres brought back from flats on to the cylinder surface. Therefore, with a closer setting at D and larger gradient between D and E, the strength of these outside air currents entering the gap at D is more. This reduces flat strip percentage.

2.14   DOFFER[1,2]

After the main carding action is over, the fibres on the cylinder wire points are carried around to the nearest points between the cylinder and doffer. At this junction, the wire surfaces on both move in the same direction (Figure 2.61). However, the wire teeth are positioned in the opposite direction. In the region A, where these wire points are closest, the setting between them is only 0.1 mm (4/1000 in).

   In comparison to the surface speed of the cylinder, (800–850 m/min or 2600–2800 ft/min) the doffer moves at a considerably slower speed of 30–35 m/min (110–120 ft/min). Therefore, as against a thin fibre film on cylinder, there is a significantly thick deposition of fibrous layer on the doffer. The process of this thick deposition on the doffer is called a condensation of fibre in the form of web. The state, in which the material becomes condensed on the doffer wires, remains exactly the

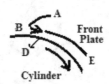

FIGURE 2.60   Control of flat strip percentage[1,2]: The flat strip contains a major proportion of lint. Many of the fibres in this strip are long enough to form normal stock. It is thus necessary to control flat strip to avoid undue lint loss.

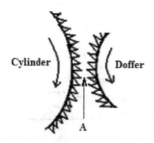

FIGURE 2.61 Closest point of approach[1,2]: The setting between the cylinder and doffer is the closest among all the card settings. The closeness helps in more efficient transfer of fibres from cylinder onto doffer.

same when it is subsequently peeled off from its surface. This material is still in the form of a thin, filmy and semi-transparent web. The high entanglement of the fibres in the web, as well as the fibre condensation in the form of thick deposition, however, helps in yielding sufficient strength to the final web coming out from the doffer.

2.14.1 FIBRE TRANSFER[1]

The fibres presented by the cylinder to the doffer are not all transferred at the same time. It was observed that with flexible clothing, only 4–5% of the fibres on the cylinder were picked-up by the doffer, whereas the rest repeatedly went around. Some of the fibres sometimes went as many as 18–20 times around the cylinder before being transferred. With metallic fillet (saw-tooth wire), however, this transfer has been significantly improved. The following are some of the points that help this fibre transfer onto the doffer.

1. As the web is continuously stripped from the doffer surface, the doffer always presents fresh and clean wire points to the oncoming fibres from the cylinder.
2. In flexible card clothing, the count of the doffer fillet was usually kept higher than that of cylinder, e.g., if the cylinder fillet count was 100^s, the doffer fillet had a 110^s count. This gave a higher number of points per unit area on the doffer, thus increasing the possibility of transfer of fibres. With metallic wires, the higher point density is not required on doffer. This is because the fibres are not held so firmly by the metallic wire on the cylinder and are easily let-off.
3. A larger diameter of cylinder presents a comparatively flatter curvature, whereas the doffer has more sharp curvature. Therefore, the fibres, once under the control of doffer wires, quickly move away from the influence of the fast moving cylinder surface.
4. The centrifugal force experienced by the fibres on the cylinder is very high and when they are near the cylinder–doffer junction this force enables them to be thrown onto the doffer. The doffer, however, runs at a comparatively slow speed and so, is unable to reciprocate by creating by creating an equal and opposite force.

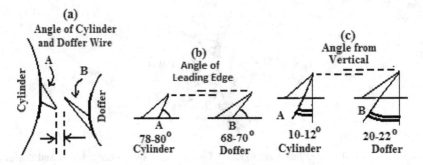

FIGURE 2.62   Angle of cylinder and doffer wires[1,4]: The higher angle of leading edge of cylinder wires allows the easy transfer of fibres onto the doffer. The comparative smaller angle of doffer wires improves their hold on the fibres and helps in better fibre retention. (a) Angle of cylinder and doffer wires, (b) angle of leading edge. (c) angle from vertical.

5. At the cylinder–doffer junction, there is a sudden release of air currents generated and carried by the former. This helps in releasing the fibres and bringing them onto the doffer.
6. When both the cylinder and doffer are at closest, the position of the teeth A (Figure 2.62a) on the cylinder and teeth B on the doffer is such that the fibres can easily come off the teeth A. Subsequently, once they are lodged on teeth B, it becomes more difficult for them to come back. Even the angles of inclination of the teeth of these two wires play an important role. The angle of the leading edge of cylinder wire (Figure 2.62b and c) is 10° to 12° (70°–80° to horizontal) whereas that of the doffer is 20° to 22° (68°–70° to horizontal). The wider angle (narrower angle with horizontal) of the doffer wire offers much better hold on the fibres once they land on it.
7. The setting between the cylinder–doffer wires is the closest of all the settings on a card. The close proximity between them helps in an easy transfer of the fibres onto the doffer.

2.14.2   Doffer Comb

The doffer comb was one of the important features of the old conventional card. As explained, the fibrous matter in the form of a thin layer deposited on the doffer was required to be subsequently peeled-off. This was done by a fast oscillating comb called a *doffer comb*.

A grooved pulley on the cylinder shaft was used to drive the eccentric shaft through a compound pulley (Figure 2.63c). The eccentric gave a rocking motion to the fork mounted on the comb shaft (Figure 2.63a).

A brass split bush (Figure 2.63b) was made to hold eccentric A. The prongs of the fork (Figure 2.63a) held this eccentric. The final rocking motion given to the comb shaft resulted in an up-and-down movement of the doffer comb blade (Figures 2.63a and c). This oscillating motion given to the doffer comb was called a *stroke of the comb*.

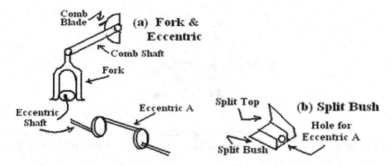

FIGURE 2.63   Driving of doffer comb[1,2]: The conventional doffer comb is given a vibratory motion. This enables it to peel the web from the doffer surface. The vibrations per unit time of the comb and the doffer surface speed must be linked. In this case, the cylinder shaft is used to drive the doffer comb pulley through ropes. (a) Fork and eccentric. (b) Split bush. Driving of doffer comb.

In very old cards, the whole assembly was enclosed in a box—called a *comb box*—and was partially submerged in oil for constant and continuous lubrication.

Even then, the brass bushes were not capable of working at very high speed. Their bearing surface used to get worn-out quickly, owing to high friction, and this considerably restricted the speeds of the doffer comb. As the strokes per minute of doffer comb were related to working speed of doffer, it was not possible to increase the doffer speed. However, with flexible wires both on the cylinder and the doffer, the productive speeds of the cards themselves were basically limited. Thus, a comb box speed of 1200 strokes per minute was quite satisfactory for a card with flexible wires. With the introduction of metallic wires on both the cylinder and the doffer, many flexible wire cards were converted to the semi-high production (SHP cards) speeds for which doffer comb speeds had to be increased. With the introduction of ball bearings and roller bearings in the comb box, it was possible to increase the strokes per minute of the comb to almost 1800 strokes per min. The roller doffing however, permitted the increase in the doffer speed up to 40 rpm (Table 2.4)

However, it became very necessary to regularly and systematically carryout the maintenance of the Comb box. The ball bearings required periodic cleaning and re-greasing. Very few mills went for this kind of scheduled maintenance for the comb box.

TABLE 2.4
Comparative Carding Speeds

No.	Speed of Organ	Flexible Card	Converted SHP Card	Earlier Generation HP Card
1.	Cylinder	165–175 rpm	220–230 rpm	300–350 rpm
2.	Licker-in	375–400 rpm	620–650 rpm	700–850 rpm
3.	Doffer	6–10 rpm	14–18 rpm	30–36
4.	Doffer Comb	1200 strokes/min	1800–2000 strokes/min	N.A.

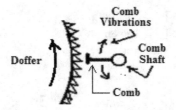

FIGURE 2.64 Position of doffer comb[1,2]: It influences the effective stroke position over the doffer. A correct positioning of the comb enables the peeling of doffer web more effectively.

On the other hand, as the doffer required continuous and effective stripping, the matching oscillating speed of the doffer comb became crucial. The relationship between the two is evident from the following example.

The movement of the doffer comb is in the arc, where A, B and C (Figure 2.65) are the three different positions during its travel from one extreme to the other. If the total movement of the comb along the arc, is (say), for example, 3.17 cm (1.25 in), then it is observed that about 35–40% of this movement is only useful in peeling-off the web. This gives only 1.27 cm (0.5 in) of the web being removed per stroke of the comb. It can be further seen that when the comb moves from B to C, it also moves a little away from the doffer, removing an additional portion of the web. Let us assume that this is 0.63 cm (0.25 in). This means the net length of the web removed will be 1.90 cm (0.75 in).

With 14 rev/min of doffer and 68.58 cm (27 in) as its diameter, the net length of the web around the doffer to be peeled-off will be 3014.77 cm (1186 in) per minute. Hence the required number of strokes of doffer comb will be (3012)/(1.90) = 1585.72 strokes per minute

An allowance for an operating efficiency of the comb can be given up to 20% further, so that the actual speed of the comb will have to be kept higher by an equal margin.

Thus, the strokes per minute for a doffer speed of 14 rpm would require about 1900 doffer–comb strokes or oscillations per minute. It was observed that, even with the ball bearings, very high comb speeds (strokes per minute) could not be reached

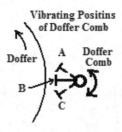

FIGURE 2.65 Peeling of doffer web[1,2]: Out of the full oscillation of doffer comb, only a part of the stroke is effective in peeling and stripping-off the web.

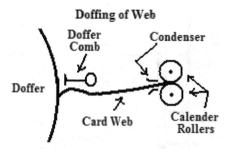

FIGURE 2.66   Calendering of web.

so easily as it led to a lot of wear and tear on the parts and the bearings used. Hence, to work within economical limits, it was necessary to restrict the doffer speed to less than 18 rpm. In high production carding, however, a revolutionary change in doffing the web from the Doffer was made in later years to reach speeds much beyond 25 rpm (see Chapter 5, Section 5.1).

In between however, a Super Uri Comb with more blades was introduced to enable the comb to strip the doffer at a faster rate. The advantage was that an oscillatory movement of doffer comb was converted to rotary motion.

However, here, too, the comb could be speeded up to 2500 strikes per minute. The comb, in this case had a device where its blades, after stripping the doffer, were retrieved. It is perhaps because of this mechanism that the wear and tear on the parts increased at higher speeds. Finally it was roller doffing which solved the problem satisfactorily. The web, after being peeled off, was led to calender rollers (Figure 2.66) for condensation.

2.15   CALENDER ROLLERS[1]

The web removed by the doffer comb is collected and compressed into sliver form. A small condenser kept before the calender roller (CR) collects the web across the width of the machine (Figure 2.66). After converging, the web is fed to the CR. The bottom CR is positively driven by a shaft carrying a CR end wheel. This wheel gets its drive from large doffer wheel through two carrier wheels (Figure 2.67). The top CR receives its drive from the bottom CR by simple gear meshing.

The speed of the calender roller (CR) is always correctly related to that of the doffer, irrespective of any change in the production rate or hank of the sliver delivered. However, there is a slight tension draft between the CR and the doffer. Usually, this tension draft serves two purposes. Firstly, it does not allow the web to sag or slacken in between the doffer and calender roller. Secondly, the pull that it exerts on the web enables its easy removal from the doffer. Nevertheless, the tension draft should be limited to a sufficiently low value so as to avoid any web stretching or any disturbance to the web. This is because the distance between the CR and doffer is much wider and the excess of draft may lead to uneven stretching of the web and

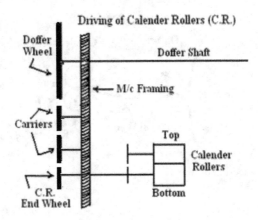

FIGURE 2.67   Drive to calender roller.

thus affecting the sliver uniformity. The CRs are weighted by additional springs (Figure 2.68) to achieve the required condensation of the gathered web.

The spring pressure can be adjusted with the help of nuts provided on either side. Thus, the pressure can be increased by tightening the nuts. This additional pressure is necessary on the top calender roller as its mere weight is not adequate enough to condense the web into a compact sliver. A clearer plate is kept in contact with the top and the bottom calender rollers. The plate has flannel cloth on its surface. Thus, any accumulation of fibrous matter or any other thing on the working surface of calender rollers is wiped-out.

2.16   COILER[1,2]

The sliver delivered by the calender rollers is led to a coiler that systematically coils it into a cylindrical can. The drive to the coiler is taken from the bottom CR wheel K_1 (Figure 2.69) which ultimately drives the bevel L_2 on the vertical coiler shaft. The

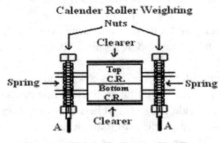

FIGURE 2.68   Spring loading on CR[1,2]: Apart from their own weight, the calender rollers are additionally weighted. This makes the sliver more compact and helps in avoiding its undue stretching in subsequent processes.

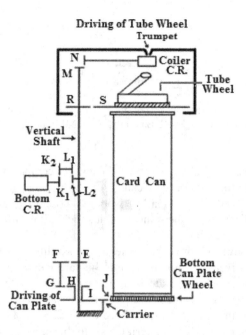

FIGURE 2.69 Drive to coiler[1,2]: The whole assembly is situated at the front side of the card. Its job is to suitably lay the coils of the slivers into the can in a systematic manner. For this, both the card can and the tube wheel are driven at different speeds.

top of the shaft carries yet another pair of bevels M and N, through which the drive is passed on to the coiler calender roller (coiler CR).

One of the coiler CRs, which gets drive through a vertical shaft, drives the other through gear connections P and Q (Figure 2.70).

The two coiler CRs are pressed against each other by means of springs. With this pressure, they are able to compress the sliver further. Finally, the sliver is led into the card can below. The surface speed of the coiler CRs is adjusted in such a way that it is slightly higher than that of calender rollers in front of doffer. This keeps the sliver between these two points in a little taut condition and does not allow it to sag or become slack. The sliver before entering the nip of two coiler CRs is passed through a small trumpet. It is very important to choose the correct size of the bore of trumpet

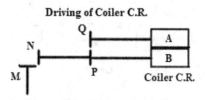

FIGURE 2.70 Top view of coiler CR gearing[2]: The two calender rollers are responsible for finally delivering the sliver in a coil form and through the tube wheel into the card can.

FIGURE 2.71   Trumpet bore[2]: The size of the bore depends on the hank of the material as well as the type of material processed. Basically, the smallest possible workable size of bore is chosen.

(Figure 2.71) because it finally determines the level of compactness imparted to the sliver during the coiling into the card can.

The bore size should be as small as workably possible to allow a satisfactory condensation of sliver, but at the same time, it should allow a free flow of material through to avoid any choking.

A satisfactory level of condensation leads to two things. Firstly, it avoids any stretching of the sliver during coiling into the card can, and secondly, it facilitates accommodation of more material into the can. This reduces a good number of creelings and piecings in the subsequent process. The use of the correct size of the bore is also important. When a bore of correct size is chosen, it helps in regulating the quantity of sliver material passing through. It is found that the uniformity of the material is best when the minimum workable diameter of trumpet hole is used. The tube wheel, positioned just below the trumpet, is driven from the vertical shaft through a pair of gear wheels R and S (Figure 2.69). It carries an inclined coiler tube (Figure 2.72) with its top end vertically below the trumpet bore and very close to the nip of coiler CR. The sliver from the nip of these rollers enters the tube and is rotated by the tube wheel. Ultimately the sliver is coiled into the card can positioned below. The upright card can is placed on a can-plate carrying a gear wheel J (Figure 2.69). The gear wheel E on the vertical shaft drives the wheel J through a train of gears—F, G, H and I (Figure 2.69). The can plate rotates at a very slow speed. The axis of the rotation of the can however, is slightly off-set from that of the tube wheel. The tube wheel lays the sliver in circular coils into the card can.

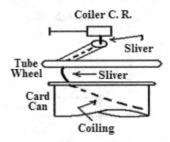

FIGURE 2.72   Coiling of sliver[2]: The sliver is passed through the inclined tube of the tube wheel. As the tube rotates, the coils of the slivers are laid into the can placed underneath.

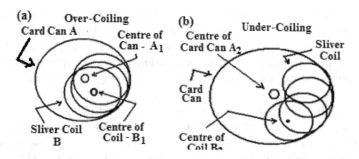

FIGURE 2.73a,b Sliver coiling[1,2]: The basic difference in these two is due to diametric size of the tube wheel in relation to the can diameter. As seen below, the diameter of the tube wheel is less than half the diameter of the can in under-coiling, whereas it is more than half the diameter in over-coiling. (a) Over-coiling and (b) under-coiling.

2.16.1 UNDER- AND OVER-COILING[3]

The diameter of the coils laid down by the tube wheel is related to diametric measurements of the card can (Figure 2.73). Depending upon the distance between the centre of the can (A_1) and the centre of the coil (B_1), the pattern of coiling is decided. In *over-coiling*, depending on the distance A_1B_1 (Figure 2.73a), the coils project a little over the centre of the can. In this case, the coil diameter is more than half the diameter of the can.

Thus, there is an empty circular space formed around the centre of the can. In the under-coiling system, an empty space is formed with the coil diameter less than half the diameter of the can. In the over-coiling system, a similar empty space is formed with the coil diameter more than half the diameter of the can (Figure 2.73b). The distance A_2B_2 (centre of can to centre of coil) is comparatively larger in the case of 'under-coiling.' As against this, the distance A_1B_1 is smaller in 'over-coiling'. These patterns decide the utilization of can capacity. Thus, utilization of can space is better in over-coiling (Figure 2.73c).

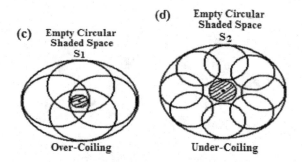

FIGURE 2.73c,d Over- and under-coiling[1,2]: In over-coiling, as the coil diameter is larger than half the diameter of the can, a smaller central hollow, thus created, gives more stability to the sliver coils laid inside the can. Also the can space utilization is also better. With under-coiling it is exactly the opposite. With large diameter of cans used in modern carding, mostly the system used is over-coiling. (c) Over-coiling and (d) under-coiling.

Also, as the coils have a wider base in over-coiling', the structure is well balanced. This is really very helpful, particularly when the bigger diameter cans are used and when the material is filled in excess over the top of the can. This is because, when these cans are doffed, the material (sliver coiled in the can) occupies a certain height over and above the top of the can. With under-coiling (Figure 2.73d), the structure is more unbalanced, so there is a tendency for the worker to tuck this over-projecting material into the can to avoid its falling down during transportation.

However, with this, the coils at the top, and therefore the sliver, are disturbed to a great extent. In the conventional system, the full cans were simply dragged over the floor. They were then taken to the next process. It used to either spoil the flooring or the inside sliver was disturbed. In better-planned mills, trolleys are provided to carry these cans to the next process. In still more modern mills, with much wider can diameters (106 cm × 106 cm / 42 in × 42 in), the cans themselves are provided with casters. With the smooth flooring (oxy-fluoride/vinyl flooring), such cans can be simply pulled/pushed to their destination. This permits their easy transportation to the next process.

REFERENCES

1. Manual of Cotton Spinning – "Carding" – W.G. Berkley, J.T. Buckley, W. Miller, G.H. Jolly, G. Batters by & F. Charley, Textile Institute, Manchester, London, Vol. 3, Butterworth, U.K. 1965
2. Elements of Cotton Spinning – Carding & Drawing - Dr. A.R. Khare, Sai Publication
3. Cotton Spinning – William Taggart
4. Process Control in Spinning – ATIRA Silver Jubilee monographs, ATIRA Publications.
5. A Practical Guide to Opening & Carding – W. Klein, Textile Institute Manual of Textile Technology, Manchester, U.K., 1987.
6. A Practical Guide to Opening & Carding – W. Klein, The Textile Institute, Manchester, U.K. 1987

3 Stripping, Burnishing and Grinding[1,2]

3.1 WIRE MAINTENANCE

The stripping, burnishing and grinding operations are required for better up-keep of the wire-clothed organs. These make the wires function more efficiently. During carding action, the fibre tufts are opened, individualized and cleaned by the action of the wire points. However, some of the fibres, along with vegetable originated impurities, get embedded into the wire clothing. The wire points themselves get worn out as they work on the cotton tufts. In the case of long stoppages and particularly in humid weather, the wires are likely to become rusted. It is thus necessary to reform the state of wires so as to make them function better. Therefore, it is very essential to keep the condition of the wires at their best so that they perform satisfactory carding action. This is only possible when a mill adopts a regular wire maintenance schedule.

3.2 STRIPPING

Though with the advent of metallic wires, the stripping operation in its true sense is not essential, it is important to occasionally clean the saw-tooth wire clothing. This is because some broken seed particles, stalks of cotton leaves or bracts get embedded between the saw teeth of adjoining wires. As these can only be removed by stripping, it is necessary to occasionally employ a stripping roller to remove the embedded matter, the period is much longer though. Therefore, unlike flexible wires, where the stripping roller is required to be employed every 4–6 hours, the stripping operation on metallic wires can, if needed, be employed, once in 20–30 days.

However, with flexible wires, the stripping operation was absolutely a must owing to the tendency of the wire clothing to absorb a far greater proportion of fibres, along with other foreign matter. It was observed that the wire clothing became substantially clogged with fibres in about 1 to 1½ hours. This made the wire points much less effective in exercising the same carding action on the material being worked. The stripping operation, therefore, had to be carried-out after each 1½-hour interval. After every stripping operation, therefore, the wires were completely cleaned and their surface was made fresh for renewed carding action.

The metallic wires may also require stripping when it is observed that there is some fibrous vegetable material clogged in the clothing. This is owing to incorrect atmospheric conditions, or damaged or worn-out wires, and is termed *lapping* or *loading* on the cylinder. This, again, is removed by employing a stripping roller. However, the duration of the stripping, in this case, has to be judiciously managed.

DOI: 10.1201/9780429486562-3

3.2.1 ROLLER STRIPPING

A wooden roller of about 15 cm (6 in) diameter and having length equal to the card width is covered with a special stripping fillet (Figure 3.1a and b). The fillet consists of flexible wires placed in a thin, pliable and soft foundation. The wires themselves are also quite soft, flexible and elastic when in action. Brackets are provided on either side of the card framing at an appropriate position (above cylinder and doffer) to carry the stripping roller. The stripping roller is put on these brackets whenever a stripping operation is required.

The long and flexible needles of the fillet of the stripping roller are set in such a way as to penetrate a little into the cylinder and doffer wires.

The needles are about 1.9 cm (0.75 in), and go inside the wires of the cylinder and doffer by about 1.6 mm to 3.1 mm (0.06 to 0.12 in).This penetration, however, should be just sufficient to effectively strip the cylinder and doffer wires, otherwise the wires of the stripping rollers, which are very soft, become damaged.

With a correct setting, the roller usually removes about a 3.8 cm (1.5 in) band of the strip from the cylinder or doffer wire surface at a time. Before the stripping operation, first the cylinder and doffer must be stopped. The feed is disconnected and the belt is shifted to a loose pulley in conventional card. The time is allowed for the cylinder to stop.

This is because, owing to inertia, it takes 2–3 minutes for the cylinder to fully stop (modern cards have a braking system).The stripping roller is placed in the brackets. The belt is shifted for a very short duration on a fast pulley and immediately returned to a loose pulley (Figure 3.2).

This allows the cylinder to rotate at a slow speed for a short duration. The grooved pulley on the stripping roller is driven by another grooved pulley which is compounded with a loose pulley on the main shaft and is cross driven by a rope. It is very

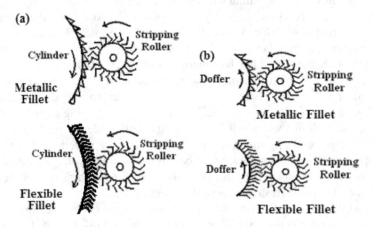

FIGURE 3.1   Wires on the cylinder & the doffer[2,3]: The wires, while working on cotton, do show a tendency to hold and accumulate a little quantity of fibres. This quantity assumes a larger proportion with the flexible wires, though with metallic wires this tendency is quite small. Apart from fibrous matter, there are also embedded impurities. The stripping becomes necessary in both these cases. (a) Stripping roller on cylinder and (b) stripping roller on doffer.

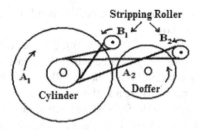

FIGURE 3.2 Driving of stripping roller[2,3]: While stripping the cylinder or doffer, it is essential to maintain the surface speed of the stripping roller much higher as compared to those of both the organs.

important that the cylinder be given only a slight rotational speed. This is because the surface speed of the stripping roller has to be much higher than that of either cylinder or doffer. Only then the required stripping action is brought about.

3.2.2 SOME OTHER TYPES OF STRIPPING MECHANISMS[3]

Vacuum stripping is done by specially shaped air suction nozzles (Figure 3.3) mounted suitably on a traverse screw shaft. The nozzle is made to traverse across the full width of cylinder or doffer to suck the fibres off the wire surface. The suction fan head and the collecting unit are centrally located, below the working surface of cylinder and doffer. A valve is provided to start or shutoff the suction.

By positioning the unit at an appropriate place, the suction can be continued even when the card is working. This avoids any loss of machine production, and as there is no direct contact with the wire surface, the life of the wire on the cylinder and doffer is increased. However, the impurities often have a tendency to go a little deeper into the wire base and hence, after a long interval, it may become necessary to employ roller stripping. The roller stripping becomes necessary at least once a week, as depositions of fine dust that settle on the wire points make them dull. The stripping roller, when employed in this case, gives a polishing effect, as well.

Continuous Stripping employs a similar type of nozzle and suction device, which is an independent unit. The nozzle fitted under the cylinder continuously sucks the

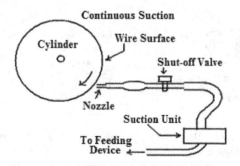

FIGURE 3.3 Vacuum stripping principle[2,3]: Roller stripping is done when necessary. However, vacuum stripping, being continuous, prolongs the working session of a card.

fibres from the underside and feeds them back beneath the lap just before the feed roller. The unit works non-stop.

Static Stripping is done by a unit that is fitted under the cylinder and forms a part of its undercasing. Static is generated when the fibres rub against air currents induced by the cylinder rotation. With the installation of this unit, the fibres are attracted towards the cylinder surface and this increases their chances of being transferred more effectively onto the doffer. Owing to this, the card works smoothly and uninterruptedly. Here again, it is still necessary to employ roller stripping once in 4–5 days as the vegetable originated impurities clogged in the foundation can never be removed by mere static action.

3.3   BURNISHING[1]

In this operation, a similar type of roller is used and it is clothed with long and widely spaced wires. However, as compared to the wires on the stripping roller, the wires on a burnishing roller are coarser, stronger and less flexible (Figure 3.4a, b and c). Also, unlike the wires of the stripping roller, they are not bent but are quite straight and hard. The hardness of the burnishing wires is greater than that of striping wires. Hence, when the burnishing roller is worked on cylinder or doffer, the roughenings or the barbs developed during working or after grinding on their wires are removed. To carry out this operation, the wires of the burnishing roller are also made to penetrate a little into the cylinder or doffer wire clothing. During the burnishing operation, the wires of the burnishing roller give a little side-polishing effect to the cylinder wires and make them sharper. This is likely to benefit the carding action.

When the burnishing roller is put on the cylinder, it is necessary to reverse its direction so that the wires of the burnishing roller work from the back side of the cylinder wires. The doffer, however, is run in the normal direction. With the introduction of metallic wires, the use of the burnishing roller lost its popularity. This is because the metallic wires are much harder than the wires of the burnishing roller, which makes the polishing action of burnishing wires far less effective.

However, the burnishing roller has been found to be very effective on flexible wires. Nonetheless, for the same reason, the penetration of these wires into those of cylinder or doffer must be carefully controlled, otherwise the cylinder wire clothing

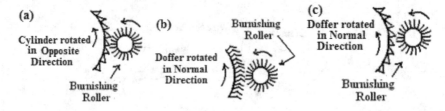

FIGURE 3.4   Burnishing roller[2,3]: While removing the burrs and rust on the wires, the operation also gives a polishing effect to the flexible and saw-tooth wire, more so to the former. It was a very common practice to apply a burnishing roller to flexible wires (normally after grinding) as only this roller could penetrate the ground wires and give them the side polishing. (a) Metallic wires on cylinder, (b) flexible wire on doffer and (c) metallic wire on doffer.

is likely to become damaged. Sometimes, manganese dioxide powder is used while applying the burnishing roller to improve its polishing action.

3.4 GRINDING[2,3]

Both, the delicate needle points in flexible clothing and the sturdy saw-tooth metallic wire points suffer heavily while carrying out the carding job. They lose their sharpness and become dull. The only difference between them is that the loss of sharpness with flexible wires occurs within 2–4 weeks (depending on the class of work), whereas the metallic wires become dull only after 7–8 months. The loss of sharpness weakens the hold of wires or saw teeth points on cotton fibres. This necessitates the grinding operation. This operation restores the good working condition of the wire points, by sharpening them and thus makes them ready for renewed and effective carding action.

The relationship between the sharpness of the cylinder wire points and their cleaning action is not well established. However, it is conceivable that as the sharper points open the tiny tufts of cotton more effectively, the resultant cleaning is going to be still better. Even then, the cylinder is not expected to specifically carryout this job.

Further, the sharpening of doffer wires also improves their ability to receive the fibres from the cylinder. The sharpened wire points thus precisely carry the fibres and do not allow them to be rolled during the fibre transportation.

When the wires become dull, apart from the reduction in their opening power, they also tend to roll the fibres. This results in formation of *neps*.

In fact, a periodic quality control check on nep level in the card web can prove to be very useful in deciding when to grind. It is well known that the metallic wire may need grinding only after 7–8 months. If the records of the weekly readings of neps per 100 square inches are maintained and plotted, it will show that at the end of a certain period, the nep level suddenly starts rising (point A – Figure 3.5). This basically depends on the quality of raw material processed and the quantity produced). It is around this time that the grinding operation should be carried out.

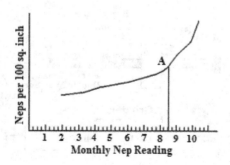

FIGURE 3.5 Periodic nep reading[2,3]: Depending upon the class of material processed, type of wire used and working hours in a week, the card wires become dull. The rate is initially slow, but later increases. Depending again on the acceptable nep level, the frequency of grinding is decided.

The flexible wires, being much softer, lose the sharpness of their points very quickly. This is because unlike the metallic wires, where the leading edge as well as a tooth-point (and later the small top plateau of the point) perform the carding action almost simultaneously, the flexible wire presents only a small, fine needle point which wears-off much faster. The frequency of the grinding operation required in their case, therefore, is once in 2–4 weeks.

3.4.1 Types of Grinding Rollers[1,2]

There are basically two types of grinding rollers: the solid or dead roller grinder, and the traversing or horse fall grinder. These are further typed as emery covered and one that uses grinding stone (made-up of stone granules). In emery covered rollers, the hard-backed emery-coated strip is clothed around the surface of the roller; the grinding stone is built-up by using small grinding-stone granules bonded with special adhesive bonds. Aluminum oxide, Zirconia alumina, Silicon carbide, etc., are some typical abrasive materials. High purity grains are manufactured with ceramic aluminum oxide. It is a very hard and strong abrasive. The grit size is another description related to the granules. It is connected with the size of the granules, the higher the number, the finer is the size of the granules.

Depending on the size of the granules or coarseness of the emery, the rollers can be classified as coarse or fine grinders. In coarser grinders, the grit size varies from 10 to 24, whereas the finer grinders have a grain size of 70 and above.

For coarse, rough and quick work, coarser grinders are used. Here the finish is not important. For fine, smooth and delicate work, finer grinders are employed. It is also possible to use a coarse grinder initially to do the major part of the work and then apply a finer grinder for a finer touch.

As the roller extends almost the full width of the cylinder or doffer, the solid grinder (Figure 3.6) is given a very short traverse. The main advantage of this roller is that it is able to even-out the whole surface across the organ quickly and simultaneously. However, the major portion of the wires under it is continuously ground and

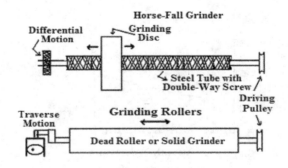

FIGURE 3.6   Types of grinders[2,3]: There are two types of grinders: solid roller grinders and horse-fall grinders. The former is in continuous contact with wires and quickly removes surface irregularities. The latter traverses the whole width of the card and is slow in its action, which evens out the surface unevenness of the wires.

thus is likely to be heated above the hardening temperature. This happens when the grinder is used over a longer period. Therefore, there is not only a danger of wire points losing their specially developed hardness, but also barbs (roughening developed on the sides of the wires due to fusing action of heat) are likely to be formed on the sides of the wires.

The traverse wheel or horse fall grinder, on the other hand, is only 8.9 cm (3½ in) wide and has a diameter of 17.8 cm (7 in). It is given a long traverse from one end of the machine to the other. The irregularities are evened-out as the grinder comes in contact with the wire surface. In this type of grinding, the wires are ground in a better way. The most important thing is that the wires are in contact with the grinder for a comparatively shorter time during each traverse. Therefore, the wires are not allowed to be heated-up to a high temperature. This gives sufficient time for the wires to cool before the grinder returns to the same portion of wires. Thus, there is less danger of *hooking* or *fusing* of wires. As compared to the emery surface, the work done by the grinder with stone granules is more powerful. The emery covered grinders were more conveniently used on flexible wires as they gave a light polishing action. They were not much used on metallic wire. This is because the metallic wires had harder surface. Normally, a stone grinder is used when grinding metallic wires.

Both the types of grinders are perfectly true (their surfaces are perfectly round and concentric, so they rotate without any eccentricity) when manufactured. But during use, the new grinder roller has to work on irregular wire surfaces. Therefore, the new grinder also experiences an uneven wear of its own surface. After extended usage, the grinder no longer remains true. The grinder then must be reshaped after a certain time interval, which depends upon how frequently it is used. A fine cutting tool may be employed for this purpose. In the case of emery surface, however, the worn-out emery clothing itself has to be replaced.

The speed of the grinder in relation to the surface that it grinds is equally important. In *slow grinding*, the speeds of the working surfaces of the cylinder and doffer are much slower than that of the grinder, whereas in *fast grinding* it is exactly the reverse (Table 3.1).

The intensity of grinding and its traversing rate (with the horse-fall grinder) are yet other factors that can be varied. When the grinder touches the wire points, sparks are produced. The length of the spark decides the intensity of grinding. Usually about a 2-inch spark length gives the normal grinding. The spark length is changed with the contact pressure between the grinder and the working surface of the wire.

TABLE 3.1

Relative Speeds of Cylinder, Doffer and Grinder

Organ	Slow Grinding		Fast Grinding	
	rpm	m/min	rpm	m/min
Cylinder	10–12	43.6	180–200	753.3
Doffer	4–6	10.7	320–340	706.5
Grinder	650–700	388.5	650–700	388.5

With a very light touch, a spark length of about 1in or less is produced. This gives a very light polishing effect on the wire points. However, when heavy irregularities of the wire need to be evened out, the contact pressure may be slightly increased. Nevertheless, the grinding period in such cases must be carefully regulated.

Normally the traversing rate of 130 in to 150 in per minute is taken as the average speed. It equals almost three traverses of grinder from one end to the other. This rate is related to the cylinder and doffer speed. With lower speeds of these organs, the traversing rate also has to be slowed. This is because the grinder is required to operate on every wire point around the whole surface of both the cylinder or the doffer and it should move a distance not more than its own width for each revolution of these two organs.

The directions of rotation of both, the working wire surface as well as the grinder, are also important. The grinder should work on the wires from the backside of the wires so that during the sharpening of the wires, their leading edge does not face the grinding action. For this, the cylinder has to be rotated in the reverse direction. The doffer, however, presents the slope of the teeth favourably in its usual working direction and hence is rotated in the normal direction during grinding.

3.4.2 SLOW AND FAST GRINDING[1]

As mentioned, in slow grinding the cylinder and doffer surface speeds are slower than that of the grinder. The driving arrangement is shown in Figure 3.7a. The loose pulley (C_3) on the cylinder is driven in the normal direction by keeping the belt running on it. This pulley is compounded with another grooved pulley (C_1) which drives another grooved pulley (A_1) of compound pulley A with a cross rope. The compound pulley is the usual pulley which drives the doffer through a barrow wheel and pinion.

Yet another grooved pulley (A_2) on this compound pulley drives the two grinders G_1 and G_2 on the cylinder and the doffer respectively. The doffer carries a flat pulley (D_1) which then drives another flat pulley (C_2) fixed on the cylinder shaft.

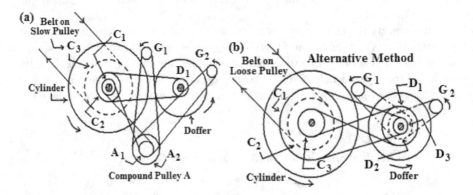

FIGURE 3.7   (a) Slow grinding and (b) fast grinding[2,3]: The terms used depending on whether the cylinder–doffer surface speeds are slower or faster than that of the grinders. In both, the surfaces of cylinder and doffer are ground. However, the class of work slightly differs.

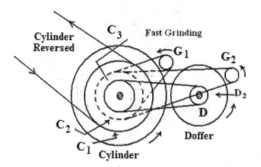

FIGURE 3.8 Fast grinding method[2,3]: The cylinder, during grinding, must be rotated in the reverse direction so that the grinder does its job from the back side of the tooth. This is necessary to protect the tip of the wires.

Thus, the cylinder is rotated from the doffer. When the doffer rotates in its normal direction and at normal slow speed, the connections mentioned above rotate the cylinder in the reverse direction and at still slower speed.

In fast grinding, there are two different methods of driving the two organs. In the first, the motor is rotated in the reverse direction (Figure 3.8) by changing the terminals in the electrical panel board. Thus, with the belt on fast pulley C_1, the cylinder is rotated in the opposite direction and at its full speed. The normal drive to the doffer is disengaged and it is run directly from cylinder (C_3 to D) at a much faster speed. The grinder is driven from the grooved pulley on the cylinder shaft. Thus, the two grinders are directly driven from the cylinder (C_2 to G_1 and G_2).

In another method (Figure 3.7b), the belt from the motor is kept on loose pulley (C_2) in the normal direction. The grooved pulley (C_3) drives the doffer pulley (D_1) with a cross rope. Here again, the normal drive to the doffer is disengaged. Another doffer pulley drives the cylinder (D_3 to C_1), thus rotating it in the opposite direction. The grinders (G_1 and G_2) are rotated from pulley (D_2) on the doffer.

3.4.3 FLAT GRINDING[1]

The normal practice in the mills is to dismount the whole flat set from the card and separately grind each on a grinding machine. The advantage here is that the wire height of each flat can be separately measured and the grinding may be done to make the heights of all individual flats the same after grinding. However, the revolving flats were ground on the machine itself. This was made possible with a specially designed device—*Higginson's* flat grinding device. During their normal movement, the direction of the wire points after the flats leave the cylinder is most favourable for applying a grinding roller. During grinding on the machine, the flats are brought to the grinding device, one after the other. However, grinding of flats on the machine or even elsewhere poses one major problem of presenting the wire surface evenly to the grinding roller. This is because while presenting the flats to the grinder, they are required to be held at their bearing surface, which has a *heel and toe* arrangement.

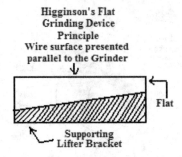

FIGURE 3.9 Provision for flat grinding[1,2]: The flats have a typical heel-and-toe arrangement for their bearing surfaces. Therefore, they could not be made to face the grinding surface evenly. Higginson's device solved this problem completely.

So a lifter bracket (Figure 3.9), having the correct shape to counterbalance the taper due to this heel-and-toe arrangement is used (Higginson's flat grinding device) to make the flat wire surface perfectly parallel to the grinder. Alternately, the flats may be taken out from the machine and then conveniently and separately ground on a flat grinding machine, which again has a similar arrangement.

It is customary to check the height of the wire points on all the flats and then record it, before and after the grinding. This is because the height of the wire-tops on all the flats must be the same after flat grinding. Whenever the flats are taken out from the machine for grinding, the mills usually have a few spare flat sets already ground and kept in reserve. This helps in reducing the stopping time of the machine and resulting production losses to a minimum.

3.4.4 PREPARATION FOR GRINDING[3]

As the grinding involves contact between the wire surface and the grinder, sparks are produced. It is therefore essential that there be no fibrous material left over on the wire clothing before carrying out grinding. This, otherwise would create fire hazards.

Before grinding the cylinder and doffer, the feed is disconnected and the lap is unrolled and kept on lap-carrying brackets. The drive to the licker-in and calender rollers is also disconnected. The machine is allowed to run idle for about an hour to ensure that all the flats get fully stripped. The machine is then stopped and stripping is carried out on cylinder and doffer. Depending on whether the grinding is slow or fast, the ropes and belts are connected. The grinding rollers are finally positioned on grinding brackets.

Initially, a small setting between the grinder and the cylinder or doffer is maintained uniformly across the width. The respective organs are then driven and slowly the grinder is brought down equally from either side. This is an operation that requires a great deal of skill and must be done with a lot of precision. It is very important to see that both the ends of the grinder are brought down equally and simultaneously. This ensures that the grinder makes contact with the wire surface evenly across the width of the machine. The end point for bringing down the grinder is decided by the length of the spark produced. This also regulates the intensity of grinding. After a short interval of time, the grinder starts uniformly missing the contact with wires across the width. This is an indication of the completion of one phase of grinding. For a light grinding, the operation usually continues for 15–25 minutes. For heavy work, the grinder is once again lowered to continue the operation. When there are greater proportions of irregularities, the grinding operation may be continued for as much as 4 to 8 hours.

REFERENCES

1. Cotton Spinning – William Taggart
2. Manual of Cotton Spinning – "Carding" – W.G. Byerley, J.T. Buckley, W. Miller, G.H. Jolley, G. Battersby & F. Charnley, Textile Institute, Butterworth Publication, 1965, Manchester
3. Elements of Cotton Spinning – Carding & Drawing – Dr. A.R. Khare, Sai Publication, 1999, Mumbai

4 Card Settings[1,2]

4.1 WHY SETTINGS?

In carding, several different parts are designed to function automatically so that there is continuous conversion of raw material from lap to sliver. The related organs are, therefore, required to be set correctly at certain close distances. The optimum setting usually depends on the type of cotton processed, the trash content in the feed material and the percentage of waste to be extracted. The mechanical condition of the machine, dynamic balancing of heavy parts rotating at high speed and the accuracy of various mountings decide the limit of their potentials in giving the required results. These settings can only be used as a guidelines, as the web quality required and the amount and kind of trash to be extracted will ultimately decide their magnitude. The closer settings, though lead to better performance of the machine, they are likely to be risky when the machine condition is poor. Apart from the fire risk, the closer settings may cause the wires to dull quickly. However, at other important places, e.g., the setting between cylinder and doffer or licker-in and cylinder, or even cylinder and flats, it may be essential to precisely carry-out the actual desired setting. Some of the settings must be done periodically, especially those which govern the trash extraction and carding action. Similarly, after every grinding operation, it is necessary to reset the relative components. The normal leaf gauges (Figure 4.1), which are ground to an accurate thickness, are used for setting the various parts.

4.2 LAP GUIDES

The setting is not critical as it merely controls the spread of the outer edges of lap. This, however, restricts the lap selvedges within the wire surface of the licker-in. But too close a setting may fold the selvedges and damage the licker-in wires, owing to a thick, folded portion of the lap suddenly entering the licker-in zone.

4.3 FEED PLATE TO LICKER-IN

The setting of 0.254 mm to 0.305 mm (10–12/1000 in) is generally used. But when there is significant change in the staple length of the fibres, an even wider setting, up to 0.43 mm (17/1000 in) is not uncommon. This is essential for avoiding any fibre breakage or damage. It is observed that a setting smaller than the one suitable for the staple length processed generally lowers the yarn strength. Too wide a setting, on the other hand, results in plucking of the cotton tufts. This leads to insufficient opening of the fibres by the licker-in prior to main carding action and ultimately affects web quality.

DOI: 10.1201/9780429486562-4

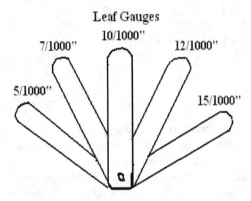

FIGURE 4.1   Leaf gauges[2,3]: Thin metal strips. They are polished and given rust-proof coating. The thickness of each strip is different and precise, and they are embossed on one side. The leaf gauges are used when setting the important organs in the card.

4.4   LICKER-IN TO CYLINDER

The licker-in is set to the cylinder at a distance of 0.127 mm (5/1000 in). This closeness is essential so that the cylinder satisfactorily removes (strips-off) the fibres from the licker-in surface. If the setting is too wide, it leads to loading of the fibres on the licker-in and neppiness in the card web.

4.5   FLATS TO CYLINDER

As the main carding action takes place between these two organs, it is important to set them correctly. Usually, the setting between them is 0.254 mm (10/1000 in); however, in some cases, the gauge is as close as 0.228 mm (9/1000 in) for better carding action. In this case, the web produced is expected to be cleaner, but the flat strip increases. Even then, the removal of neps is always better with a closer setting.

With bulky fibres (polyester, polyamide and acrylic), it is necessary to widen the setting between cylinder and flat (0.38 mm to 0.43 mm, 15/1000 into 17/1000 in), otherwise the fibres are treated too aggressively. In the case of cotton, however, too wide a setting results in poor carding action and inefficient nep removal. This leads to a poor appearance of the card web.

4.6   DOFFER TO CYLINDER

This is a very important setting and should be adjusted very accurately across the width. The gauge is 0.127 mm (5/1000 in), but when the wire surfaces are ground to perfect uniformity and when the parts are mounted with precision and are balanced perfectly, a still closer setting of 0.101 mm (4/1000 in) is possible.

A closer setting allows the doffer to effectively carry away the fibres from the cylinder. For higher production (through coarser hank), the setting may be slightly widened. However, this should be carefully observed, as a too wide setting results in overloading of the cylinder. This is because the fibres on the cylinder are not transferred efficiently. As a result, the carding action suffers and the web delivered by the doffer becomes patchy and cloudy.

4.7 DOFFER COMB TO DOFFER AND FLAT COMB TO FLATS

As both the combs travel in a circular arc, they must be set at their closest point of approach with the respective surfaces. In the case of doffer comb, the arrangement is comparatively simple and set screws are provided to vary the length of the rod between their fulcrum and a point where the rod holds the comb.

Though both the settings are not critical, they should be adjusted carefully and the distance should be close enough to remove the fibrous material from the respective organs over which they work. Apart from the setting, the condition of the comb blade is also important. The teeth of the comb, if damaged, should be polished or repaired from time to time. Care must also be taken to see that the comb teeth do not touch the wires over which they are required to work.

(Settings around the licker-in region, setting of the back plate, front plate and cylinder and licker-in undercasing were discussed earlier, and are not repeated here.)

REFERENCES

1. Cotton Spinning – William Taggart
2. Manual of Cotton Spinning – "Carding" – W.G. Byerley, J.T. Buckley, W. Miller, G.H. Jolley, G. Battersby & F. Charnley, Textile Institute, Butterworth Publication, 1965, Manchester
3. Elements of Cotton Spinning – Carding & Drawing – Dr. A.R. Khare, Sai Publication 1999, Mumbai

5 Developments in Carding

5.1 DIRECTION OF DEVELOPMENTS

The carding process has two main objectives: individualization of fibres and extracting the remaining trash from the lap. A lot of research work has been done in carding. The efforts are aimed at improving either one or both the functions. With the advent of high production carding technology, the concept of *Shirley pressure points* was introduced and now has been incorporated on all modern cards. The concept involves controlling the air currents, streamlining them and avoiding generation of undue pressure points within a card. This is done to improve both the objectives cited above. Therefore, apart from improving trash extraction, it also tries to boost the opening power of the card. When increasing the card production, it is very important that trash extraction and individualization of fibres should never be put at stake at any time.

The suction system attached to the card on the basis of pressure points basically avoids fly, fluff and fine dust being liberated into the card room atmosphere. It also avoids the development of high pressure points. In this way, it helps in saving the lint and improving the carding action.

5.2 PRINCIPLES OF HIGH PRODUCTION CARDING[7]

In conventional carding, the carding action deteriorates when the production rate is increased. From a technological point of view, the licker-in region and cylinder–flat region are some of the places where suitable modifications can increase the intensity of carding, essential for higher production rates. This is necessary to ensure that there is no loss of quality when working the card at a higher production rate. In the cylinder–flat region, the intensity of carding can be increased by decreasing the fibre load on the cylinder. This can be done by reducing the layers of fibres between them to a workable minimum. These layers can be imagined to be divided into different sub-layers, e.g., (a) the fibres embedded in the cylinder wire foundation, (b) the operational layer, consisting of the quantity of fibres the cylinder acquires during steady state and (c) the fibres on the surface of the flats.

With the introduction of metallic wires, the first sub-layer, viz. the fibres embedded in the cylinder wire foundation, has been reduced to an insignificant amount. In a way, this increased the productivity by almost eliminating the stripping and reducing the frequency of grinding (prevalent with flexible wires). The second layer, called the *operational layer*, allows a certain proportion of fibres to be transferred onto the doffer during every revolution of the cylinder. When the setting and other parameters are kept constant, there are basically two factors that affect the build-up of fibres in this layer on the cylinder: (a) production rate and (b) proportion of fibres transferred onto the doffer.

DOI: 10.1201/9780429486562-5

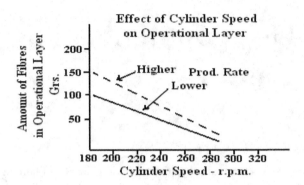

FIGURE 5.1 Operational layer on cylinder.[2,7] The layers carry the store of fibres on the cylinder from which the fibres are transferred on to the doffer from time to time. The efforts are always directed to reduce the quantity of fibres (loading) in this layer.

When the production rate is kept constant and the cylinder speed is increased, the second factor, i.e., the proportion of fibres transferred onto the doffer, increases owing to higher centrifugal force experienced by the fibres during transfer. Obviously, the amount of fibres in the operational layer therefore decreases. Thus, it is essential that whenever higher production rates are intended, the cylinder speed has to be increased so as to counteract an increase in the fibres in the operational layer.

It can be seen from the graph (Figure 5.1) that though the amount of fibres in the operational layer is more for higher production rate, the difference between the two (one with higher production rate and the other with lower production rate) decreases with increase in cylinder speed.

The amount of fibres in the operational layer also depends on the hank of sliver. With finer hank (less weight per unit length), the fibres in the operational layer are reduced, thus inducing better carding action. The profile of the cylinder and doffer wires and the setting between them also plays an important role in reducing load on the cylinder. The design of the cylinder wire teeth is such that it provides a strong carding power, and at the same time, allows the fibres to be easily transferred onto the doffer. On the other hand, the doffer wires are able to entrap all the fibres presented to them. For this, the height of the doffer wire is more than that of those on the cylinder. In addition to this, the angle of teeth of the doffer wire at the base is much less compared to that of the cylinder. Owing to wider base angle (carding angle) of the cylinder wires, more wire points can be accommodated and this increases carding power.

There is yet another sub-layer which consists of the fibres on the surface of the flats. At a constant production rate, when the flat speed is increased, the flat strip increases. However, the quantity of fibres under each flat decreases considerably. Hence, the flat speed has to be also increased at higher production rate so as to keep this third sub-layer to a minimum. It may be noted here that the increase in the amount of flat strip is proportional to the square root of increased flat speed. Therefore, in order to counteract any increase in the flat strip owing to higher flat speed, new types of flat tops (Cresta Diamond, Eureka, Supra, Picco tops, etc.) are now available. They are semi-rigid or fully rigid–type flat tops and present hardened sharp points for better carding action. They accumulate much less quantity of flat strip but retain vegetable-originated impurities very well.

In order to improve carding action, it is necessary that the size of the tufts presented to the cylinder is very small. This has been achieved by intensifying licker-in action. Apart from increasing its speed, other modifications around the licker-in have also been introduced. Some manufacturers have modified this region by having two or three licker-in rollers or putting a pair of toothed rollers below the licker-in. In some other cards, better opening is achieved by having a comb bar below the feed plate and closer to the licker-in. To improve the openness of the material and extraction of trash, modern cards invariably have *carding segments*. With them, not only the trash but also the tuft size presented to the cylinder–flat action is very much reduced.

Increasing doffer speed is yet another way of increasing production rate. Incidentally, higher doffer speeds also reduce loading on the cylinder. The main limitation for increasing doffer speed in conventional cards, however, was the speed of doffer comb. With some improvement, it was possible to work this comb at a maximum speed of 2500 strokes per minute. Even at this speed, the maximum workable doffer speed was around 20–21 rpm. This is far less than the expected and the desired speed thought for high production rate. The roller doffing was the first successful step and the concept was introduced by Platt to obviate this limitation. It was most popularly is known as the Crosrol Varga unit and had totally replaced the doffer comb. The unit was thus installed on what was then Platt's high production card.

5.2.1 Crosrol Varga[2,4]

In this attachment, a stripping roller is used to peel the web from the doffer (Figure 5.2). This roller is clothed with special wires and set very close to the doffer. This, together with the special shape of the wire points, helps in very effectively removing the fibre layer on the doffer.

Owing to the shape of wires, the stripping roll not only strips the fibrous web from the doffer but also easily lets it off to the next roller, called the *redirecting roller*, again set close to the stripping roller. The stripping mechanism to peel the web from

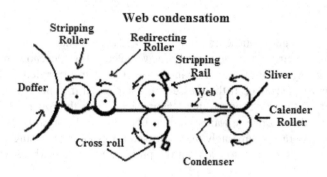

FIGURE 5.2 Web condensation with Crosrol Varga unit[2,8]: Pioneering efforts were made to use a circular wire-clothed roller to doff the doffer web. A successful implementation of this idea removed the main mechanical limitation of increasing the doffer speed.

the doffer was incorporated for the first time by introducing a roller-doffing device in place of vibrating comb.

With the introduction of the stripping roller, it was possible to increase the speed of doffer-stripping operation quite appreciably. This enabled a substantial increase in the doffer speed and led to high production carding. However, in practice, the working doffer speeds were initially only in the range of 35–40 rpm. This led to about a 75–100% increase over the doffer speed of converted metallic card. Limitations, however, came from another angle—the cylinder speed. With converted metallic wire card, it was not possible to speed-up heavy and giant cylinders beyond 280 rpm. The cylinder shaft in the old card was not strong and sturdy, as it was not made of steel. With iron shafts, it was not possible to accurately and correctly balance the heavy cylinder, either statically or dynamically, especially at higher speeds. A strong and sturdy steel shaft was required to hold the heavy cylinder, the increase in load on the shaft being due to the weight of metallic wires. There were also problems in conveying the doffed web through stripping and redirecting rollers to the crosrol and finally to calender rollers. These were because of lapping problems around the Crosrol Varga unit. In addition, the crosrols (or crush rolls) in the Varga unit were heavily weighted. The heavy pressure led to crushing of the remains of small broken seed particles, forming a sticky surface. The solution was thus not in carding but back in the blow room. It was essential to see that no broken seed particles remained in the material fed to the card.

The Crosrol Varga unit incorporated three exclusive features which were vital to successful application of high production technique in the card.

- Even distribution of effective crushing pressure along the full width of rolls with a self-adjusting *off-setting* arrangement. The crushing pressure was infinitely variable between 2 lb/in and 12 lb/in along the card width.
- Combined with stripping, a precisely controlled draft was applied between crosrols and doffer for absolute control over the carded web.
- A controlled draft was also applied to the web between the calender rollers and the cross rolls.

The crosrols are highly polished, smooth and plated heavy rollers (approx. 40–45 kg). In addition to their own weight, they are weighted by an eccentric mechanism. However, the ingenuity of this mechanism was that it incorporated an *off-setting* arrangement of the rollers. If the pressure was applied at the end of the two cylindrical rollers with their axes parallel (Figure 5.3a), both the rollers would deflect at the centre, the top one upwards and the bottom one downwards. With off-setting, the rollers are very minutely made crossed (exaggerated crossing view shown in Figure 5.3-b). This tries to even-out the pressure across the whole width of the rollers.

In consequence, there would be an excess of pressure at the end but comparatively much less pressure at the centre. Even if the rollers were cambered to barrel shape, the specific contour would only be applicable to a given calculated pressure. For the

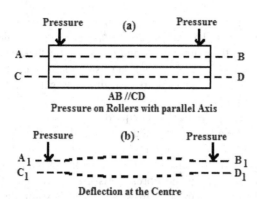

FIGURE 5.3 Pressure on parallel surfaces.

pressure higher or lower than this, the crushing of impurities would not be uniform across the width of the card. It was, however, essential that for optimum crushing efficiency, the pressure on the cross roll should be uniform along its length. This was done by off-setting the rollers (making their axes finitely non-parallel), in proportion to the applied pressure (Figure 5.4).

The mechanism of off-setting is automatically arrived at, and the operator has only to adjust the required pressure by turning the eccentric. The scale is marked to enable the correct stepping in arriving at either increasing or decreasing pressure. A compound cam or eccentric simultaneously sets both the pressure and the degree of off-setting. There is also a provision to lift the rollers so that there is a very minute gap between them. This is done when the card is being ground or burnished to avoid fretting corrosion. A zero gauging may be used when processing synthetic fibres, as they do not need any crushing action of the crosrols.

The rubbing of crushing rolls against the web leads to generation of static charges, which are passed on to the fibres processed. This is more severe in cold climates when the humidity is low. Moreover, the seed bits may get crushed and the oil, along with the waxy matter, may try to adhere to the crushing roll. It is, therefore, essential to apply slightly higher draft between the crush roll and doffer, and the calender roller and crush rolls, especially in the latter region.

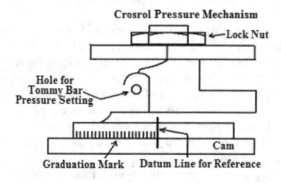

FIGURE 5.4 Pressure on Crosrol Varga.

Owing to the high pressure of crush rolls, however, the dry vegetable-originated impurities are pulverized. The higher draft imposed between calender rollers and crush rolls loosens-up these impurities and gives a little shaking to the web. This causes the pulverized matter to fall down in this region. Even the neps can get flattened and develop a greater surface area, so that they get easily picked up in the subsequent combing process.

Along with higher production, it is equally important to maintain quality. In this context, it may be again mentioned here that the higher production through higher doffer speed is preferred to one through coarse hank. This is because the higher doffer speed offers increased fibre-collecting surface and therefore the power to more easily pick-up the fibres. This greatly reduces the fibre-load on the cylinder. With coarser hank of card sliver, there is slight deterioration in the web quality.

5.2.2 Fluff Generation

High production carding also poses the problems of fly and fluff generation. This is not only due to higher working speeds, but also due to the non-holding type of wire fillet (metallic wires have much less tendency to hold the fibres). Shirley introduced the *Shirley Pressure Points* system to counteract this. The system can be adapted to fit to either a single card or a group of cards linked together.

The filter box is mounted on one side of the machine and is usually put on separate framing. The suction unit is thus separately located. The ductings are made to connect the suction unit to various different regions: (1) between licker-in and cylinder, (2) on the back side of the working flats and (3) between doffer and cylinder (below front plate), etc.

The solid Plastic pipes with slits for directing the suction are positioned in these regions over the respective surfaces and are connected to the filter box through rubber tubing. Thus, the fly and fluff generated in these areas are gathered and collected over a screen in the filter box. A small window is provided at the bottom for the operator to open and remove the gathered fly and fluff at regular intervals. The periodic removal of gathered fly, fluff and dust is important, as it cleans the screen and enables the suction unit to function normally, more effectively and continuously. On a modern card, the fly and fluff do not accumulate over the screen. They are conveniently taken away and separately collected.

There is another objective of providing suction. The regions where the suction nozzles are made to reach are the high pressure points. By providing suction in these regions, the pressure is reduced and this helps in streamlining the air currents around the cylinder and other areas. Even at normal carding speeds, the pressure points are developed. Basically, the development of these pressure points is owing to the air currents being made to suddenly pass through a narrow gap (setting) between some of these organs. This is further aggravated when the parts are moving in opposite directions. At high production speeds, the pressure generated is much higher and it leads to develop very strong air currents in these regions. Especially at cylinder–licker-in and cylinder–doffer junction points, there is a pumping effect owing to air currents trying to force their way through such narrow gaps. When suction is provided, it helps to reduce this pressure and avoid any turbulence which otherwise would certainly disturb the carding action.

5.2.3 Necessity of Stop Motions and Slow Speed Drive to Doffer[2]

Whether at low production or high production levels, the stop motions are very necessary, as they protect the machine from mechanical damage and also avoid wastage of material. The stop motions should be able to detect the presence of any foreign body, especially metal particles, to avoid expensive damage to the card clothing. A double lap feed is sensed by the stop motion and the feed to the card is immediately stopped. In the absence of such motion, the thicker material trying to pass through the narrow setting regions is very likely to damage the clothing or break the parts. Especially with double lap feed, a sudden impact experienced by either mote knives or licker-in wires is likely to lead to serious and heavy damage.

With old, conventional cards, the sliver breaks at the front used to be mechanically sensed by counter-weighted silver spoon guides. In modern cards, the absence of sliver is sensed electronically and the message is relayed though electrical connections to immediately stop the feed and doffer. The lap run-out, web break at calender roller, lapping at calender roller and choking at the coiler trumpet are all sensed electronically to cutoff the feed and stop the doffer.

With these stop motions, not only the working performance of the machine is improved but also the sliver waste is reduced quite significantly. The fault indicator lamps glow whenever there is a fault. This helps in quickly drawing the attention of the tenter. A yellow lamp usually indicates the sliver break, the red one conveys the lap-ups, the green one points to full-can doff and the white one denotes the normal running of the machine. Special provisions are made to enable the worker to stop the machine from either the front or back side of the machine. It is possible for the worker to de-activate the sliver break stop motion temporarily during piecing-up of the web through the coiler tube and coiler calender roller.

It must be remembered that maintaining the stop motions in top condition helps in three ways: firstly, it ensures machine safety; secondly, it reduces production of soft waste; and thirdly, it improves the operative efficiency. In high production units, the last factor is very important, as in ultra-modern mills, there are fewer persons to mind the machines and the operator is also assigned several other related duties. With an efficient stop motion coupled with indicator lamps, it becomes easy for the operator to quickly mend or repair the fault and restart the machine. This helps in keeping the down-time to an absolute minimum. Even the production of soft waste, which is substantially reduced owing to efficient stop motions, is an added advantage. This is because the soft waste must be recycled. This invites more disturbances in the product and forces compromises in the quality.

With higher doffer speeds, it becomes increasingly difficult to piece-up the web. It therefore becomes essential that, during this operation, the doffer should revolve at a comparatively slower speed to facilitate easy mending of broken web or sliver. After the piecing-up operation, the normal higher doffer speed can be easily restored. On some of the latest modern cards, this change-over is carried out electronically and steplessly so as to give a steady increase or decrease in doffer speed over a predetermined period.

It is thus possible to increase the production of the modern card by 300–400% over that of the conventional one. The points which are discussed above invariably become the feature of any modern card of today. The mills going for higher production have to incorporate the modern developments. In this connection, the replacement of old card with the modern high speed and high production card provides the financial justification, too. Though the initial investment in the modern card of today is high (more than 4–5 times), the saving in labour, space and power, along with improved layout and better card room conditions, can more than compensate for the initial cost in the short run (shorter pay-back period). Quality-wise, also, the sliver is far more clean and free from neps to give better yarn.

5.3   INDIA-ROLL[2,4]

The higher trash content in Indian cottons is partly attributed to the ginning conditions prevalent in the country. The Indian cottons thus present some peculiar problems to the spinners. In addition to the higher amount of trash, the presence of broken seed particles poses serious problems, especially when these particles are crushed while passing through the crosrols. With poor blow-room performance, the seed fragments are easily passed on through the lap-feed material to the card. In addition to this, another problem that affects the functioning of crosrol is the humidity, especially in the monsoon season. A practice was usually followed in many mills to reduce the pressure on the crush rolls to overcome this problem. Also, the card was worked at a slower speed to avoid frequent lapping around the crush rolls. In such cases, the advantages of both the web purifying action of the crosrol and higher production rates are lost.

Apart from contaminating the cotton stock being processed, the sticky surface of the crush rollers often led to the lapping of fibre web around crosrols as the result of oil oozing from the crushed seed particles. Scraper blades were provided to lightly touch the surface of the crosrols. But they were able to remove only solid matter sticking to the surface, so the oil coating persisted and became a source of trouble for the web sticking on crosrols.

Similarly, with high humidity, the high pressure (8.7 kg/cm along the length of the roller) on the crush rolls expelled moisture and little of the wax from the fibres and made the rolls wet. This again led to web sticking. With these two problems, it became very difficult to work Indian cottons on the card at higher production rates with normal crush-roll pressure.

In India-Roll, the web is stripped from the doffer, as usual, by a stripping roller (Figure 5.5), and then directly picked up by the offset spring-loaded crush rolls for the web purification. A second stripping roller, clothed with similar type of wires as that of the first one, then strips the crush roll and passes the web more positively onto a pair of polished guide rollers. Subsequently, the web is condensed as it passes through a trumpet. A pair of calender rollers finally delivers the sliver.

The teeth of the second stripping roller are made to work in close proximity to the surface of the crush rolls; so, the web sticking tendency, according to the claims of the manufacturers, is considerably reduced.

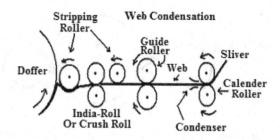

FIGURE 5.5 India-Roll[2,8]: This is a modified version of the Crosrol Varga. The seed bits and particles often were crushed under the pressure of Crosrol, the oil oozing out in the process making the surface of the crosrol sticky. An additional stripping roller provided immediately after the India-Roll partly solved this problem.

5.4 TANDEM CARDING[5]

The lap usually fed to the card consists of a thick sheet of cotton tufts. Even with the best possible maintenance, the natural variations present in the lap result in uneven distribution of the fibres on cylinder and flats. This leads to loading of fibres on their surfaces. Owing to poor control over this distribution, the fibres are not carded effectively. The concept of double carding had emerged as a solution to these problems. But it involved making a feeding lap from the sliver of the first card. However, this difficulty was later solved and came the concept of *tandem carding*.

The system is essentially a double carding process (Figure 5.6) where the web is processed twice without an intermediate stage of forming sliver. The lap, after undergoing the first carding process, is subjected again to the second carding by designing a special transfer mechanism. During this, the material is automatically transferred from one card (first section) to the other (second section). Earlier, when double carding was used (even in woollen and waste spinning, the concept of breaker and finisher card was involved), the sliver from the first card was converted to a suitable sheet form (Derby Doubler principle) and then fed to the second card. This process required re-condensation, which reduced the effectiveness of the second carding process.

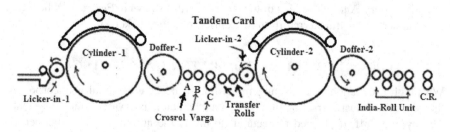

FIGURE 5.6 Tandem carding[2,8]: The concept of carding the material twice has its roots in wool carding. However, with cotton the first carding delivered the material not in lap form but in sliver form. Hence it required another lap preparation system. When the two cards were suitably joined, making a lap twice was not necessary.

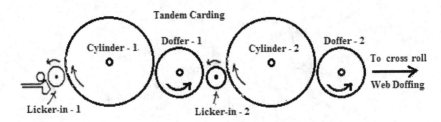

FIGURE 5.7 Tandem carding[2,8]: In this process, the main focus is on how the web is transferred from first carding operation to the second one. As also shown in Figure 5.6, this transfer can be done in two ways: (a) the first doffer to the licker-in of the second card section through cross roll unit and (b) the first doffer to the licker-in of the second card directly.

In tandem carding, however, the web from the first carding section is presented in the same web form to the second carding section, which is also a part of the same machine. Hence, the level of individualization already achieved in the first carding section is not only maintained but further enhanced during second stage. The ideal level of fibre-to-fibre separation is thus expected to be reached in this type of card. Further, the cleaning action, owing to double carding, is almost complete. On an average, the sliver from a tandem card is much cleaner and therefore superior.

A tandem card can be looked upon as made up of two high-production cards suitably joined together with a small difference in design which helps in transferring the web from the first card to the second one.

The stripping roller A removes the web from the doffer of the first card and then redirecting roller B assists in feeding this web to a pair of crush rolls C. On emerging from the nip of crush rolls, the web is carried further with the help of two transfer rollers. These rollers carry the web over their top surface so that the quality of the web produced by the first card can be inspected (through a transparent screen). Finally, this thin web is fed to the licker-in of the second card. The thin web then passing through the second card receives another thorough carding treatment. The fibre entanglement and matted layers of fibres, if any, are freed completely from the seed particles and the trash. The remaining impurities that escape the action of earlier card are subsequently fully pulverized through India-Roll. The pulverized matter, owing to its less adhesive power, usually falls down between India-Roll and calender rollers. Another version of the tandem card is shown in Figure 5.7 where the material is directly transferred from first doffer to second licker-in.

5.5 DUST AND FLY EXTRACTION IN TANDEM CARDING

With high production rates of tandem carding, it became essential to incorporate the centralized dust extraction unit as well as the automatic waste removal system. The fly and fluff, if allowed to accumulate inside the normal bottom chamber of this card, were very likely to contaminate the out-going sliver. Aerodynamically designed plenums (spaces) are provided at strategic positions on the card to provide optimum extraction performance in the areas of dust and fly origins. This avoids any accumulation of fly in the interspaces of the carding machine. With their efficient removal, the sliver cleanliness is further improved. It was a common experience that yarns made by using tandem carding fetched a higher price in the market.

In the initial stages of tandem card, the trash dropping under the licker-in region piled-up very fast. Also, owing to the then-existing manual fly collection system, the trash withdrawal became irregular and uneven. It was thus essential to take care of piled-up trash. This is because, if not removed, this trash would otherwise have gone back into working material. Therefore, it became essential to prepare a special pit under the high production card (on the licker-in side) so that its depth under the licker-in was approximately 0.6 to 0.8 m; on the doffer side, it was 5–10 cm. A taper was thus obtained from delivery to feed side. This allowed sufficient volume of space under the licker-in.

With tandem carding, very high production rates of 35–45 kg/h were possible with cleaning efficiency levels between 90 and 95%, owing to which even lower grade varieties yielded a much better quality of clean yarn. In some instances, with better grade cottons, the quality of the material produced was comparable to combed varieties. It was also possible to save the comber noil with the use of tandem card. With the action of two cards combined, the quality of the fibre mixing was also further improved and it substantially reduced faults like streaky yarns.

5.6   DOFF MASTER[2]

When the flexible card was converted to a metallic card, semi-high production rates with doffer speeds up to 18 rpm were made possible. However, such levels of production were only possible with medium and lower grade mixings. Even then, as mentioned earlier, the doffer comb speed required to meet the doffer speed was quite high. With a speed beyond 2100 strokes/min, the wear and tear of the comb-box was alarmingly high. In spite of improved maintenance, it was not possible to cope-up with the higher comb-box speed. Also, the cost of maintenance for running the comb boxes at such high speeds increased disproportionately. Any effort in saving the cost of maintaining the comb-boxes resulted in their frequent breakdown.

The introduction of doff master was a step before the introduction of the actual roller doffing principle adopted in high production carding. Doff master was again a roller, but with flutings and not with clothed wires. It proved to be quite effective at the speeds with which converted metallic clothing card worked. The fluted roller was set close enough to the doffer at a distance of 0.7 to 0.75 mm (25/1000 into 27/1000 in). The doff master (or the roller) was made to work in the opposite direction to the rotating surface of the doffer (respective arrows in Figure 5.8).

Plate A and its setting with the doffer were key factors in allowing successful removal of the web by doff master. Plate B was kept close to the bottom side of the doff master to support the web being peeled away from the doffer. The flannel-covered clearer roller and the wire brush were positioned at the top so as to keep the surface of the doff master clean.

The action of the doff master was based on three things:

• The closeness of its setting with doffer
• The direction of its rotation
• Setting of plate A

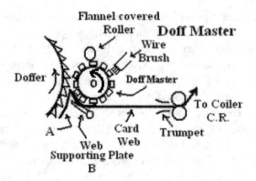

FIGURE 5.8 Doff master[2]: This was an economical alternative to the doffer comb, to speed up the doffer. However, unlike normal cross roll units, this did not have a web crushing device, which improved the web cleanliness.

It was felt that the stripping action of the doff master on the doffer was not very positive. In the mills, some trials were carried-out by replacing the doffer comb by a doff master (roller).

The doffer was run again in the same range of 18–20 rpm or at slightly higher speed. The significant advantage then was that heavy expenses for the purchase of a comb box and its maintenance could be eliminated. Compared to this, the cost of doff master equipment was quite low and hardly required any special running maintenance other than normal lubrication.

5.7 MODIFIED LICKER-IN REGION[2,4]

5.7.1 FIBRE RETRIEVER

The droppings under the licker-in are known to contain a proportion of lint ranging from 25–30%. As compared to flat strip, though it is quite low, the sizable amount of spinnable fibres may prove to be more costly, especially when processing fine and superfine cottons. As it is, with these cottons, the trash content is also very low; so, a significant proportion of lint in the licker-in dropping would mean a loss of valuable fibres. Many modifications around the licker-in region have been developed to control fibre loss.

The *fibre retriever* is, in fact, a very simple arrangement consisting of only two tin sheets—A and B (Figure 5.9). The former is hinged to the underside of the feed plate, while the latter is supported at its two sides by the licker-in undercasing. The high rotational speed of the licker-in sets-up strong air currents around it. By partially opening the waste removal hinge door at C, a feeble but definite air supply is ensured.

This air curves around the plate A and moves upwards. The plate B helps in directing this feeble air to follow the main stream of licker-in air currents. In fact, plate B, extending right up to the bottom floor, discourages the feeble air from going anywhere but upwards. As mentioned, the strength of this feeble air subsequently following the licker-in air currents is comparatively weak and it does not prevent, in any way, the normally liberated impurities, freed from lap, from falling down. However, these feeble air currents quite effectively prevent the lighter lint from

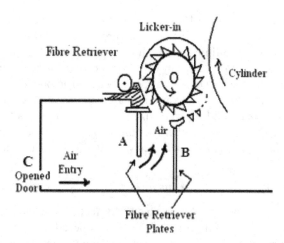

FIGURE 5.9 Fibre retriever[2]: The fibre retriever was among several efforts to save the lint proportion in licker-in droppings. It was economical and required a deflector plate in place of the normal mote knives.

falling down. In fact, they support the lint and send it back to follow the air currents around the licker-in. The lint, thus supported, enters the gap between the licker-in and its undercasing and is saved.

5.7.2   HI-DOME

Hi-Domes (Figure 5.10a–c) are the modified covers over the licker-in. It is observed that a relatively high pressure is developed in the region above the licker-in. This is because at a point immediately past licker-in/cylinder junction (A), the air currents which escape being carried by the cylinder try to enter the gap between the licker-in and its top cover. The gap A is, however, very small so a high pressure region is developed. This adversely affects the lint and trash separation at the licker-in undercasing owing to development of back pressure in this region.

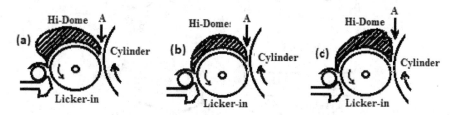

FIGURE 5.10   Different Hi-Dome shapes[2]: There is often a congestion of air currents owing to closer settings between cylinder and licker-in. The rushing air currents entering the licker-in top cover region cause air turbulence. The back pressure thus developed finally leads to more droppings under the licker-in. Hi-dome allows larger space over the licker-in and tries to calm this disturbance.

With a suitable design of Hi-Dome, a large volume is provided at the top of the licker-in and this helps in allowing the pressure to drop down in the region. As a result, it helps in streamlining the high pressure area at a point A and further minimizes the tendency to develop any back pressure at the licker-in undercasing. As shown in Figure 5.10, the Hi-Dome is a specially designed bonnet over the licker-in in place of a normal cover. Several variations are possible (Figure 5.10a–c).

The research work carried-out in connection with the use of a Hi-Dome clearly shows that, though the quantity of licker-in droppings is not much affected, the proportion of trash in the dropping is increased with significant reduction in lint content.

5.7.3   DEFLECTOR PLATE AND MODIFIED LICKER-IN UNDERCASING REGION

A *deflector plate* is a mild steel strip about 3/16 inches thick and extends the full width of the card. The working edge of the plate is bevelled at 45° to give a sharp edge, similar to mote knives. The plate is provided with slots (Figure 5.11A) cut at intervals along its length. The slots are used to fix the plate by bolts to the bottom part of the feed plate. The setting of the plate with the licker-in is done at its working edge. However, this setting is comparatively wider than that of mote knives.

In their modification, the Shirley Institute, Manchester (UK), redesigned their licker-in undercasing by making a shorter grid section of 17.5 cm (7 in) with three bars in place of two in the older undercasing. The success of the undercasing of this type mainly depends on the smoothness of the nose.

Any roughness, even to a small extent, developed around the nose area disturbs the air currents passing through the gap between licker-in and undercasing. In addition, the rough areas may start picking-up the fibres, which at times, may also lead to choking in the region. The operating efficiency is thus severely impaired. The setting at the nose is 6.35 mm (¼ in).

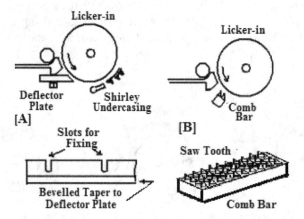

FIGURE 5.11   Comb bar and deflector plate[2,15]: In place of conventional mote knives, the use of a deflector plate improved the trash extraction and also controlled the lint loss under the licker -in. The comb bar improved the opening of fibre tufts. Thus, with both, the cleaning efficiency at licker-in was improved.

A positive gradient is maintained from the nose to the end of the sieve portion so that an uninterrupted smooth fibre flow is preserved. It is usually observed that, when the licker-in speed is higher than 600 rpm, the proportion of droppings under the licker-in increases. Further, the droppings are richer in lint. Therefore, if the proportion of the lint in the dropping cannot be controlled by widening the nose setting, a longer undercasing (longer length from the nose to the end of sieve portion) needs to be used. The deflector plate is used with a modified undercasing so as to improve cleaning efficiency with lower lint loss at the licker-in zone.

5.8 NEPS IN CARDING[9]

Neps are tangled masses of fibres. They appear as specks in white cloth and as uneven dyed spots in coloured fabrics. They tend to lower the yarn strength and increase the end breaks in spinning. The tangling or rolling over of the fibres during the process leads to formation of neps[9] and the phenomena is basically due to loss of control over the fibre movement.

A cotton with a low maturity level is highly susceptible to nep formation. This is because the immature or dead fibres tend to easily roll or curl around when subjected to beating or rubbing action. Thus, the maturity level of the raw material greatly influences the appearance of neps in the card web—the more the immaturity of the material, the more neps in the card web. The fibres which form the neps are essentially shorter in nature. In certain situations, however, higher neppiness may also mean heavy fibre breakage. This is because neps may also be formed when processing longer-staple fibres which are under strain during processing. However, with man-made fibres, where there is no element of maturity, neps are still seen in the card web.

It has been observed that the nep level in carding increases when its production rates are increased. However, even at the same doffer speed, when the feed is increased (coarser hank of lap), the neps in the card web are found to increase. On the other hand, when the sliver hank is made lighter, the nep level falls down. A well set card with good mechanical condition should remove more than 50% of the neps presented to it. When the conditions are adverse, a card may also create neps. The most important influence on this is the condition of the card wires; if the wires are not sharp, they lead to high nep levels in the card web.

5.8.1 EFFECT OF CARDING PARAMETERS ON NEPS

1. A proper synchronization between licker-in and cylinder speed not only reduces nep formation but also permits higher production rates. With an increase only in licker-in speed, there are:

   - Few fibres per licker-in tooth
   - Better opening and cleaning of the material processed
   - Uniform distribution of fibres on cylinder

   ATIRA has defined the *degree of treatment* for the action of the licker-in. This degree of treatment is influenced both by the number of wire points passing through the fringe of lap per unit time and the velocity at which they do so. However, the degree of treatment seems to have only a limited effect on the neps in the card sliver.

2. The speed of the flats is important only when flexible tops are used. A higher flat speed is reported to have better nep removal. However, with strip-less or metallic flat tops, the flat speed hardly seems to matter. The closer setting of the flats to the cylinder avoids any rolling of fibres and this is likely to reduce neps.

3. The closer setting between doffer and cylinder is conducive to better transfer of fibres on the doffer. This leads to less reworking of fibres around the cylinder and reduces nep formation.

4. When the doffer speed is increased at a constant production rate, the neps content in the web is reduced. This is because, in this case, the doffer strips the cylinder more effectively making more fresh wire points available for the incoming fibres from the cylinder. Even when card production is increased by increasing the overall speed of the machine, the neps do not seem to increase significantly. However, when much higher production rates than the normal are tried by simply increasing the doffer speed, the neppiness of the sliver as well as that of the yarn increases. This is also accompanied by a fall in the yarn strength and more end breaks at ring spinning.

5. The neps are highly correlated to cylinder loading where the fibres start accumulating on the cylinder wire surface. For a given production rate, the rate of cylinder loading increases rapidly when the feed is increased. With excessive humidity and damaged wires, the cylinder loading becomes a serious problem. Especially when the cylinder wires get contaminated with oil licking from shattered seeds or with waxy material present on the fibre surface, or even "honeydew", its surface becomes sticky. All these factors lead to more cylinder-loading and leads to more nep formation. The loading on the doffer is comparatively very small and does not seem to affect the nep formation much.

6. With flexible wire, the stripping frequency had a profound influence on the formation of neps. However, with metallic card, the stripping frequency does not much influence the nep generation.

7. Use of split tops or reversing the direction of flat has been reported to give an advantage in controlling nep level.

8. The frequency of grinding the cylinder–doffer wires is perhaps the most important criterion governing nep level. Dull and rough wire points give the rolling action to the fibres. This is purely because of inadequate control over the fibres by wire points. The need for grinding of wires on these two organs, and especially the cylinder, can be judged correctly if regular quality control checks on the neps in the card web are observed. The nep level in the card web is such a good measure of the condition of the wire points that, if the nep readings before and after grinding do not show significant difference, it clearly indicates improper grinding. Even the grinding of flats, is usually neglected. Following its correct schedule, is very important, as it also influences the nep formation.

The popularity that the tandem card received initially had the same context of significant nep reduction. With double carding incorporated in this machine, the appearance of the web was very close to that of combed material. A spinner who wanted to comb the material purely for appearance and not much for the strength of the yarn could think of omitting the combing process, if the material was processed on tandem carding. This also gave an added advantage—saving lint which otherwise would have gone as comber noil. The yarn from the tandem card is thus cleaner, brighter, almost nep-free and more regular.

5.9   FIBRE HOOKS

Detailed research work by Morton[11] revealed that the bulk of the fibres leaving the doffer in the form of web have their ends hooked (Figure 5.12). About 50–55% of the hooks are in the trailing direction; 20–25% of fibres have hooks in leading direction. These are often referred to as majority of the hooks or major hooks (trailing) and minority of the hooks or minor hooks (leading). It is also known that hook extent A_1 (Figure 5.13) in the case of trailing hooks is comparatively more than that of leading hooks A_2.

Ideally, the yarn-forming process requires that all fibres be well parallelized, aligned along the axis of the main strand and also quite straightened before they are twisted together. The hooking of fibres reduces the total fibre length available along the axis of the strand.

Thus, the fibre length available is only B_1 or B_2 (Figure 5.13) instead of the actual length of the fibre as $(A_1 + B_1)$ or $(A_2 + B_2)$. Increasing the cylinder and doffer speed has profound influence on the proportion of the hooks. With higher cylinder speed, the trailing hooks are known to increase; whereas, with higher doffer speeds, the proportion of leading hooks increases. The yarn quality in the ring frame largely depends on both the direction and the number of major hooks in the fed material. Hence, their reduction in the process prior to spinning and subsequent to carding has great importance in deciding both the quality and performance of the yarn in final spinning.

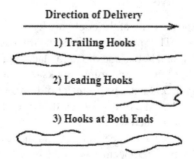

FIGURE 5.12   Fibre hooks in carding.

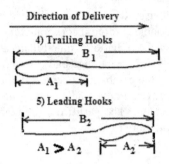

FIGURE 5.13   Fibre extent and hook extent.

The hooks get straightened during every drafting operation. In combing, owing to the very nature of the process itself, there is appreciable amount of fibre straightening. The cylinder needles in combing, while passing through the lap fringe, straighten-out the leading hooks. However, some straightening of trailing hook does occur when the top comb enters the fringe.

In drafting, the trailing hooks are favourably straightened out. It has been shown that the configuration of the fibres in the lap fed to the card does not materially affect the pattern of hook formation in the card sliver. This means that irrespective of whether or not the fibres in the lap have hooks, the fibres in the card web have a fairly typical and stable hooking pattern. The action of the flats does not materially alter the pattern of hooking. The cottons that have low fibre friction and high recovery from compression are transferred more easily from the cylinder onto the doffer and consequently the percentage of hooks formed is reduced. The cottons with low micronaire (finer) are carded easily, thus reducing hook formation. The negative rake wire, usually used for man-made and synthetic fibres to improve transfer efficiency, reduces nep level in viscose, but when used for cotton, does not assist, in any way, to reducing the total hooks formed in the card web.

5.9.1   FORMATION OF HOOKS[1]

According to Morton[11], the majority of the hooks are leading on the cylinder, whereas they are trailing in the web. The speed of the flats and their spacing over the cylinder influence the formation of total hooks and the ratio of leading to trailing hooks on the cylinder. Thus, the source of major and minor hooks owes its origin to the action of cylinder and flats. In Figure 5.14a and b, a transfer of fibres with leading and trailing hooks is shown. Also, for the sake of clarity, the distance between the cylinder and the doffer is purposely drawn much wider.

The arrow on the fibre is for reference and it shows the reversal or non-reversal of the fibres. The reversal of the fibre occurs when the doffer wires catch the rear end of the fibres (Figure 5.14a and b, 1 and 3).

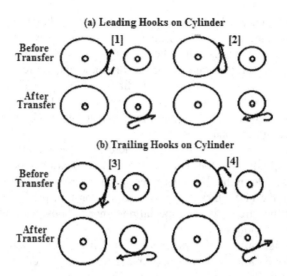

FIGURE 5.14 Formation of hooks[2,11]: The fibres, on cylinder are expected to have majority of hooks. The speed of the flats and their spacing over the cylinder influence the formation of total hooks. The rear ends of the fibres are straightened as they pass through flats. The phenomenon of the rear or front end of the fibre being caught by the doffer decides whether there is reversal or no reversal of the hooks. (a) Leading hooks on cylinder. (b) Trailing hooks on cylinder.

On the cylinder, a majority of fibres have leading hooks. This phenomenon occurs because the fibres are carried by the cylinder wires, their rear ends straightened owing to their journey through the cylinder–flat zone. The rear ends of the fibres thus usually project out and are held by the doffer wires. This induces fibre reversal. However, when the leading end of the fibres is engaged by the doffer, there is no reversal and the fibres are transferred as-is.

The reversing is caused mainly by the whipping action of the fast moving cylinder. During this, the fibre ends carried by the cylinder are initially straightened. The whipping action also involves straining the fibres, and when the fibre end is suddenly released while being transferred onto the doffer, the fibre springs back, owing to its elasticity. This also causes the fibre ends to curl or hook. Therefore, the directional effect of major (trailing) and minor (leading) hooks in the card sliver is caused firstly by the action of cylinder–flats and secondly by the doffer, because there is an alteration of the pattern of hooking during the transfer. This is accomplished through a complex phenomenon of reversal and non-reversal of fibres. Some of the fibres which are hooked may even come out as straight fibres and vice-versa.

5.9.2 Tracer Fibre Technique

Morton and others[12] were the first to use dyed fibres (non-isotopic) as the tracers in describing the classical fibre arrangement in the card sliver. Basically, this involves mixing a small proportion of such dyed fibres in a lap fed to the card. For studying yarn structure of man-made filaments, one coloured filament is introduced into a bundle of white ones. The sliver (even roving or yarn) thus produced is held in clamps and immersed in a liquid having a refractive index very nearly the same as that of non-coloured material. When the dyed fibres are viewed through an ordinary microscope or micro-projector under a monochromatic light source, the dyed (tracer) fibres stand out clearly against the white fibres which give a dim background. The fibres can then be traced suitably or recorded on a film. In another method, these fibres are viewed from two different positions, one at right angles to the other, so as to obtain a three-dimensional view.

Later, the tracer fibres were dyed with fluorescent dye and mixed by spreading them over the lap. The web thus collected on a black board can be observed in the dark under ultraviolet light. The fluorescent dyed fibres glow distinctly, and when viewed through the glass, they can also be traced on transparent paper. The fibre arrangement in the card sliver with special reference to hook pattern has been studied using these techniques.

Radioactive isotopic tracers are very useful in process control. These emit three types of radiation: (a) particles, (b) particles with high energy electrons and (c) particles having high penetrating power. The emissions owing to radiation can be detected and traced by: (1) gas-filled detection (Geiger counter), (2) scintillation and (3) photographic emulsion. For a speedy and continuous recording, isotopic tracers can be very effective.

The transfer of the fibres through carding has been studied with the advent of the radioactive tracer technique. It was found that a fibre, on average, goes 18 times around the cylinder before it is transferred onto the doffer. Thus, the tracer fibre technique seems to be very useful, versatile and sensitive enough to detect the fibre movement, its configuration and its alignment during the process. With radioactive isotopes, the problems in migration and blending have also been studied. Even the fibre movement in drafting and their acceleration can be recorded with utmost precision.

5.10   TRANSFER EFFICIENCY

The fibres presented by the cylinder to the doffer are not all transferred at the same time. It has been observed that very few fibres are picked-up by the doffer; the rest go back around the cylinder for as many as 18–20 times. Krylov[14] stated that only 20% of the fibres are removed by the doffer in their first passage around the cylinder. The longer the fibre remains on the cylinder surface, the smaller the probability that it will be removed by the doffer. Nevertheless, as stated by Krylov[14], the probability of the fibre removed by the doffer remains finite for at least 200 revolutions of the cylinder. It is interesting to note that this probability (say, after 100 revolutions), no doubt exists. Especially when the fibres on the cylinder surface are taken by the doffer, these deep lying fibres (almost closer to the cylinder bare surface) are peeled along with those taken by the doffer. If not peeled, they are at least significantly disturbed. Such disturbed fibres are very likely to be picked up by the doffer in their subsequent passage around the cylinder.

The basic principle in calculating the transfer efficiency is to find the proportion of fibres transferred from the cylinder onto the doffer in one revolution of the cylinder. By carrying-out a simple experiment (on flexible card), transfer efficiency can be easily calculated.

5.10.1 Method 1

A card is run with the feed roller feeding the lap. After 5 minutes, the feed is disengaged. During these 5 minutes, the material is collected at the doffer and weighed ('p' g). The card is stopped. The cylinder is then stripped. The strippings are weighed ('w' g).

$$\text{Transfer Efficiency} = (p \times 100) / \left(w + p\right)$$

The drawback of this method is that the flat strip is neglected.

5.10.2 Method 2

A card is run for 5 minutes with the feed on. During this time, the material in the form of web is allowed to be dropped down. After 5 minutes, the feed is disengaged and simultaneously the doffer is stopped. The cylinder, however, is allowed to run for 1 minute.

Only the doffer is then started. The first one-foot length of the sliver is collected and weighed ('p' g). After a foot-length, the total material delivered by the doffer (feed still disconnected) is collected and weighed ('w' g).

The length taken up by the doffer $= \pi \times$ doffer dia. $\times$ doffer rev. in 1 min

$$= \pi \times d \times n \left\lceil \text{doffer dia. in feet} \right\rceil \tag{5.1}$$

If 1 foot of material weighs 'p' g, then weight of the material of length in (5.1) would weigh

$$= \pi \times d \times n \times pg$$

With cylinder rpm as 'N', the material delivered by the cylinder (or material transferred onto the doffer) would be $= [\pi \times d \times n \times p]/Ng$

The transfer efficiency is the weight transferred on to the doffer per revolution of the cylinder.

Hence,

$$\textbf{Transfer Efficiency} = \frac{\pi \times d \times n \times p}{N\left(w + p\right)}$$

5.10.3 Factors Related to Transfer Efficiency

1. With an increase in production rate, it is found that transfer efficiency is higher. However, when the production rates are achieved through higher doffer speed, keeping the cylinder speed constant, loading on the cylinder is greater and it reflects on poor carding quality. However, when the cylinder speed itself is increased, there is a decrease in fibre load on the cylinder

wires. In fact, increasing the cylinder speed decreases this load on all the sections of the carding surfaces, including flats.

2. When the flats are set a little wider, there is improvement in transfer effi-ciency. However, this is mainly due to poor opening of the tufts. The unsat-isfactory opening of the tuft results in the fibre layers remaining in a raised condition on the cylinder wire surface and this helps the doffer to easily pickup these tufts.

3. When overall cylinder speed is increased, it also increases the centrifugal force experienced by the fibres on the cylinder, thus easing their transfer onto the doffer.

4. The setting between the doffer and the cylinder directly controls the trans-fer. Thus, a closer setting between them improves the transfer of fibres.

5. Perforated doffers with pneumatic suction directed at the cylinder–doffer junction point can be tried to encourage the transfer.

6. Poor transfer efficiency is definitely undesirable. This is because it leads to higher cylinder loading and results in overworking of the fibres. The fibres are taken round and round the cylinder for a greater number of times which leads to higher nep generation.

7. The ideal angle of cylinder wire inclination would be 90°. However, this angle might not give satisfactory carding action. The angle of cylinder wire (from horizontal it is usually kept between 78° and 82°). The wire angle of the doffer is generally kept narrow (around 60°) so that it does not allow the fibres, once entrapped, to go back onto the cylinder.

8. At a constant production rate, when the doffer speed is increased, it is neces-sary to reduce the hank of the sliver. Higher doffer speed offers more fresh wire surface for the fibres coming from the cylinder to land on the doffer. As a corollary, the load on the cylinder is reduced and this improves the carding.

Thus, it can be deduced that heavier sliver increases cylinder loading and reduces transfer efficiency. It means that transfer efficiency would be always higher with finer hank of the card sliver. The higher wire point density on the doffer definitely helps the transfer of fibres. Even then, the effect of the respective wire angles is definitely more pronounced.

5.11 PROCESSING OF MAN-MADE FIBRES AND THEIR BLENDS[10]

Whereas, the type of synthetic and man-made fibre to be blended with any given cotton or other natural fibre is decided on the basis of its end-use, there is a certain practical limitation on the length of the man-made fibres that can be processed on cotton processing machinery. The length of the fibre, in this case, should not exceed 40–45 mm (1-9/16 in) for blending.

Man-made or synthetic fibres require a different set-up of machinery conditions and they also have different standards for the extraction of waste allowed during the processing. Thus, they require different parameters in the blow room and card-ing. Waste must be minimized as much as possible to avoid a fibre losses. Unlike cotton, man-made fibres do not contain trash or impurities. The blending of cotton with man-made fibres, in the majority of cases, is done after the two components are separately carded. However, there are some mills which carry blow room blending for better mixing of the component fibres. The card could also be considered as a good mixer, owing to its potential for individualizing the fibres—a prerequisite for intimate blending. But the carding conditions for cotton and man-made fibres differ significantly, so a preference is given to post-card blending.

5.11.1 Prerequisites

Owing to the high bulk and loading characteristics of these fibres, the laps fed to the card are required to be much lighter in weight as compared to those made for cotton. A lower lap weight in the region of 300–350 g/m is usually preferred. The lap licking is also a serious problem with such fibres. This must be adequately tackled in the blow room, otherwise the resulting thicker portion due to the double laps usually cause damage to the wires during the carding operation. The chute feeding is not common. However, if and when used, it could totally obviate the problem of lap licking.

Tinting is done in the blow room to identify and differentiate between different blends, their percentage or the types of fibres. It helps in segregating the blends in subsequent processes.

5.11.2 Carding of Man-made and Synthetic Fibres

Viscose staple fibre is easily carded on a conventional cotton card with only a few adjustments. In conventional carding, the licker-in speed has to be reduced to 350–400 rpm, whereas the flat speed (with conventional tops) should be in the range of 2.5–3.0 cm (1 into 1.14 in) per minute. The negative rake wire must be used on the licker-in to enable quick and easy transfer of fibres onto the cylinder. This avoids possible damage to the fibres due to overbeating. The licker-in undercasing is usually covered so as to avoid waste in the form of licker-in droppings. The research carried-out in early days revealed that the bottom mote knife played an important role in this case. Based on this, some of the old mills, while processing man-made and synthetic fibres used only one knife (bottom knife) to avoid excessive fibre loss under the licker-n. Even then, apart from this, a thorough opening in the licker-in region is very likely to extract a few melted & fused fibres.

On Rieter's modern card (C-70), an adjustable mote knife for extracting normal trash and impurities is provided. However, when processing man-made or synthetic fibres, this knife (Figure 5.15) is closed. The setting between the feed plate and the licker-in must be widened to avoid any damage to the longer length of these fibres. Instead, the shape of the feed plate, ideally suited for longer lengths, can be used. Any fibre damage at this point lowers the yarn strength and results in higher end breaks in the ring spinning. The setting of 0.38–0.43 mm (15–17/1000 in) at the feed plate proves to be beneficial.

The cylinder to doffer setting should be as close as possible. This ensures better transfer of fibres and prevents the fibres from going around the cylinder repeatedly, thus avoiding overbeating. A higher bulk of man-made fibres (lower specific gravity, except for viscose) necessitates a wider setting between cylinder and flats; otherwise

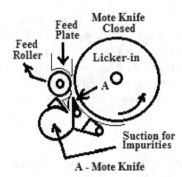

A - Mote Knife

FIGURE 5.15 Adjustable mote knife[2,15]: In modern card, some manufacturers provide a unique arrangement of adjusting the mote knife concentrically. With this, it is possible to maximize trash extraction and minimize lint loss.

the carding action is too harsh. Polyester fibre presents greater difficulties with flexible-wire fillet, particularly when the finer denier (< 1.2 den) fibres are processed. Here the problem of cylinder loading is often faced. With metallic wires, however, the cylinder loading problem is considerably less. It is also advisable to reduce the production rates; the coarser denier fibres do permit a little higher production rate, though. On a converted metallic, semi-high production card, doffer speeds ranging from 12 to 15 rpm were common, whereas with modern high production card, speeds around 25 rpm or even slightly higher are being successfully used. Owing to higher bulk, it is also customary to produce lighter sliver with 3.0 to 3.3 g/m (around 50 grs/yd—0.166 Ne).

Eureka types of flat tops with the chisel type of sharp metallic points are very useful. They provide strong carding action and at the same time extract much less flat strip. The normal flat speed of 8 to 11 cm/min (3.5–4.5 in/min) can be used. However, some mills use still lower speeds to further reduce the flat strip. It is observed that a higher surface speed ratio of cylinder to licker-in is beneficial for carding synthetic fibres, especially when processing finer denier fibres.

Adding soft waste in blow room blending should be strictly avoided. Apart from the effect of other blow room parameters, the lap licking tendency for polyester at card may be also due to excessive spin finish and ant-static finish. The artificial crimping given to the fibres during manufacture must also meet the official standards (usually 5–6 crimps/cm). The depth of the crimp is also another criterion. The shallowness of the crimp may lead to a lap-licking tendency.

The choking of sliver in the trumpet is another typical problem associated with man-made fibres. Their specific gravity being lower than that of cotton (except viscose which is a man-made fibre) the sliver becomes bulkier. Therefore, the trumpet bore size has to be a little wider when processing man-made fibres. So also, the whole passage through the coiler tube needs to be smooth and highly polished to avoid any stretching arising out of friction and possible static generation..

A control on both the temperature and humidity is essential. The relative humidity between 52% and 55% with a dry bulb between 85°F and 90°F gives satisfactory results. Some imported lots of polyester fibre bales have slightly higher moisture content. In such cases, it is always advisable to dry them first by properly conditioning.

However, during this time, care should be taken to check the level of spin-finish and anti-static finish (0.15–0.2%) after conditioning. The spin finish serves a very important function during man-made fibre processing. Apart from being anti-static in nature, it reduces friction among the fibres as well as that between fibre and metal. And there is another side to the spin finish. When combined with dust, it becomes hard and forms a coating on metal surfaces. Especially in carding, the clothing is attacked by such deposits. If the spin finish penetrates the drafting rollers or aprons, it makes them swell or sometimes even crack.

These deposits have an attacking effect on the base metal and make the metal surface sticky. This obstructs a free flow of fibres. Stainless steel is used to overcome this problem. For cylinder–doffer covers, high precision aluminium is used. These parts are anodized to counter the destructive effects of man-made fibres. Compared with cotton card, where three pre-openers are essential, the man-made card has a single roller, but it is covered with special needling. The needles are quite strong and sturdy, and are effective in improving opening efficiency.

5.11.2.1 Types of Wires for Man-Made and Synthetic Fibres[2]

Synthetic fibres are very sensitive to action of machine parameters and hence in carding, there needs to be an appropriate and suitable selection of card wires to avoid possibility of any fibre damage. In general, the characteristics of fibre processed, production rates and speeds of the organs (licker-in or cylinder), influence the selection of wire for processing man-made fibres.

Though the positive wire on licker-in can provide a better opening action, it also has more retention tendency. The negative rake wire (0° to –10°), is comparatively gentle in its action and gives better transfer of fibres onto the cylinder. The wire particulars for processing finer fibres may need small changes. Thus, with a 5°–10° angle (-ve rake) and around 80–100 points/in^2, the licker-in speed (modern card) would be higher for coarser fibre (> 1.0 dtex) and would be in the range of 1200–1400 rpm, whereas for finer fibres (less than 0.6 dtex) it would be in the range of 900–1000 rpm.

Similarly, for cylinder speeds of 700 – 800 and wire angles of 20° to 30° (from vertical), the wire point density would be around 650 points/in^2 for coarser fibre. For finer fibres, only point density needs to be higher (750 points/in^2). With the doffer, it is not necessary to vary the particulars. Both the point density and angle almost remain the same (around 400 points/in^2 and wire angle around 35°). For flats, only the point density need be changed (for coarser fibre, around 450 points/in^2; for finer fibre, around 550 points/in^2).

5.12 CHUTE FEED TO CARD AND AUTO-LEVELLING[2,15]

A step towards decisive rationalization in carding is to feed the card directly from blow room flocks—usually called *flock-feeding* or *chute feeding* (Figure 5.16). In the past, blow room production rates were very high compared to then-existing cards. The number of carding machines that a blow room could cater was therefore large. For example, with a conventional flexible or converted metallic wire card, 25–40 cards were required to balance the blow room production. It was not practical to

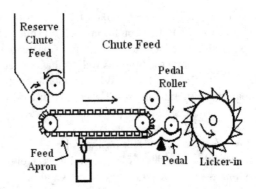

FIGURE 5.16   Chute feeding to card (earlier generation)[2,15]: Instead of lap feed, in modern carding, well opened loose fibres are directly fed. However, an arrangement needs to be made to provide this loose fibrous material in a somewhat loose and fluffy sheet form.

link the blow room material directly to so many cards. The chute feeding system, however, has gained its present status because of much higher production rates now prevailing in carding. With a real high production card, there are hardly 4–8 cards required to balance the production of a modern blow room with chute feeding.

An essential requirement for the success of a chute feeding system is to avoid too much diversity of mixings and frequent mixing changes. In most Indian mills, however, there are many mixings run in the blow room. Further, the order for a sale yarn also fluctuates often, and this necessitates very frequent changes in the day-to-day blow room mixings. The chute feed system really cannot be worked more profitably in these situations. With very few mixings, however, it is perhaps a bold and positive step towards modernization, automation and rationalization, though it may involve higher capital investment.

5.12.1   Automated Fibre Distribution System to Cards[2]

The distribution system involves the fibre flock being pneumatically conveyed from a central source to a series of card stations. Simple electrical controls from each individual card relay the signals to the central feeding system when the feed to the card unit is either started or stopped, depending on the need. An air dissipation box is mounted over each carding station and it accepts the air–fibre mixture from the source. The fibres are fed to the card using air to carry them from the source. The stock is subsequently released through a duct opening into a suitable chamber conveniently placed in the department. The filters purify the air before releasing it into the atmosphere. The design of the box is such that the air velocity inside the box drops considerably and allows the fibres to fall freely from a stream. Each *air dissipation box* is equipped with a photoelectric level control, and when the fibre level becomes low at any station, the control restarts the flow to the unit. The system is very simple and does not require any special wiring. The duct is so designed so that they do not give any problem of return air. Further, the recirculation of fibres within the duct is totally avoided. The power requirement and the noise level are quite low.

5.12.2 Concept of an Automated System[2,4]

At one time, few manufacturers embraced the idea of an automated system, i.e., automatically moving the material from blow room to card. In Platt's automated system (Figure 5.17a), the blow room, carding and drawing were all linked up in one process so that the cotton fed from bales was delivered in the form of slivers and was directly coiled into draw frame cans. This linking is possible only up to the draw frame, as the production level up to this point can be conveniently and easily balanced. In subsequent processes, however, the level of production per unit is low, as in fly frame and ring frame each spindle forms the production unit.

For satisfactory operation at the final spinning frame, adequate parallelization, proper individualization and correct direction of fibre presentation to the ring frame must be maintained. In Platt's automated system, the card sliver is directly led to Mercury drawing frame. Thereafter, only one passage of speed frame is needed prior

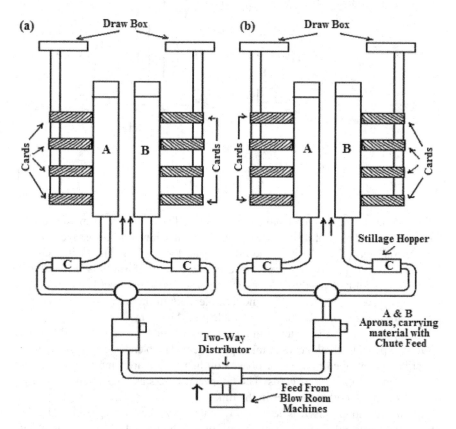

FIGURE 5.17 Automated system[2,4]: Blow room to drawing: At one time, blow room, carding and drawing were separate departments. This required more space and intensive labour. Automation to transform bale directly to a drawn sliver was a boon to the textile industry, though the initial capital expenses were quite high. (a) Platt's automated system – blow room to drawing. (b) Platt's automated card room line. *(Continued)*

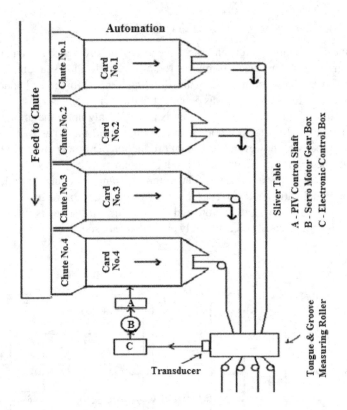

FIGURE 5.17 *(Continued)*

to ring frame. In the conventional process, where the laps were prepared in blow room, the multiple doublings of card room sliver in draw frame passages used to improve long term irregularities and compensate for any inadequacy in blending. With a very short card room process however, the sliver uniformity is required to be maintained and it becomes very necessary to have highly efficient blending system in the early processing stages. The modern blenders in the blow room do take the utmost care in achieving this objective of adequate blending.

In Platt's system[2,4], the material in the last part of the blow room is gathered in a *stillage* hopper. The final blending takes place here. The blended stock is then fed to the card at a controlled rate. The problems of distribution are simplified by making a single hopper to feed four cards. The hopper also acts as a reservoir to ensure a constant supply of cotton to cards in the link.

For this, the hopper itself is equipped with a reserve feed arrangement. The final delivery, however, takes place through the chute, which has two oscillating sheets, one at the front and the other at the back. They ensure that the material delivered from the bottom is in the controlled form. Also, if the card is stopped for some reason, it does not allow the material to become more compact in the chute. This again helps in maintaining uniform density in the chute.

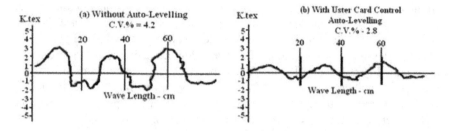

FIGURE 5.18   Auto-levelling: A draw box situated in front of the card is equipped with this mechanism. Here the mass of material entering the draw box is continuously measured and deviations are corrected by adjusting the draft in the draw box. This is done automatically. (a) Without auto-levelling (C.V.% = 4.2), (b) with Uster card control auto-levelling (C.V.% = 2.8).

To provide an identical feed to each card, a common conveyor apron carries the stock, and the material is swept into four chutes simultaneously. Also, to ensure a constant level of material in the chute, a wiper roller is provided to sweep the surplus material into an air conveying system, which carries it back into the stillage hopper. In order to ensure a constant output, the drive to all four doffers in sequence is electrically controlled. The feed system of each card, however, is mechanically linked to its doffer in the normal way.

The slivers from four cards pass along a table positioned at the front of the cards. The slivers are thus laid side by side on the table (Figure 5.17b) and are delivered to the back of the drawbox.

These four slivers are passed through an auto-leveller to ensure a regular and uniform sliver output. The device consists of a pair of rollers, the top one with a tongue and the bottom one appropriately grooved. They are placed at the entry of the draw box.

The sliver, while passing through a pair of these rollers, is assessed for its thickness and the corresponding displacement of the roller is conveyed to the transducer. The total output of the group of cards is then controlled by varying the rate of feed to one card, using infinitely variable gears operated electrically by servo-motors.

The auto-levelling system on a card can control the count variation significantly. However, this does not influence the yarn regularity directly. The Uster card controller (UCC) regulates the short and long term variations (Figure 5.18). Such a system is very useful for production of a uniform card sliver intended for direct spinning on an open end rotor spinning machine. UCC corrects all irregularities longer than 10 cm sliver length and so provides an ideal condition for constant count consistency.

Auto levelling at card corrects long-term variations, whereas that on the draw box corrects short term variation. This type of system can control variations from 30% to 50% at the card and ±15% at the draw box.

5.13   CARD MAINTENANCE[2]

A running card must always give its best performance. When newly installed, perhaps, optimizing important parameters can enable the card to operate at maximum efficiency. However, when the same card is worked over a number of years, only an optimum can be expected from the same machine. And this is only possible when card maintenance has been regularly and systematically carried out. Hence, for optimum performance, card maintenance becomes very important. It is slightly more difficult to lay down the conditions for scheduling the maintenance procedure, as this usually differs from mill to mill and depends on the constraints existing in the mills. The condition of machines, the availability and willingness of the mill to spend on spares, and the type of raw material processed—all greatly influence the cycle of operations. Similarly, the outlook of management as regards rational utilization of both men and machine also becomes very important.

Each machine, during the course of its working life, needs some periodic attention with respect to cleanliness, lubrication, rectification of disturbed settings and timings and replacement of faulty parts. The added need for inspection and replacement of worn-out parts, bearings and wheels, falls under preventive maintenance. In some cases, however, a sudden and premature failure of mechanisms must also be attended to. This is break-down maintenance. Proper preventive maintenance will reduce the frequency and occurrence of such break-downs. The preventive maintenance not only ensures trouble-free running of the machine for a longer time but also helps in reducing store consumption.

Apart from smooth working of the machine, preventive maintenance in carding is closely related to the quality performance of the machine. The full and half setting procedures must be carried-out after defined intervals; otherwise they can seriously affect the cleaning at licker-in and the quality of the card sliver and may even lead to high lint loss. The grinding procedure in the case of the cylinder, flats and doffer wires, and polishing of the licker-in wires, when carried out in a timely fashion, can greatly improve the quality of the web produced.

With increases in the costs of labour, the number of staff engaged in carrying-out maintenance is usually reduced to a minimum. However, attention must be paid to ensure a good standard of work. The excess staff cannot be easily retrenched. In many cases only reallocation is possible. Whenever maintenance activities are carefully organized and reviewed, it is possible to reduce the labour complement required. Often the reorganization of maintenance labour helps in reducing machine stoppages. This is because reorganizing activities can reduce the time required to carry them out. As mentioned earlier, the timely preventive maintenance considerably reduces the occurrence of premature break-downs.

With high-speed machinery, machine utilization is an important factor because the breakdown of a machine leads to great production loss. With higher production rates per machine, fewer machines are required so the breakdown of any one of them can easily upset the balance of production. General guidelines are given in Table 5.1; however, the actual schedule can be suitably modified depending on the existing conditions in the mill. It may be noted here that some of the procedures (*) are automated and electronically carried-out in high production card adopting 'User's Friendly' screen.

As indicated in items 21 to 23, the quality control studies must be taken in coordination with the setting and grinding schedules. On typical modern cards, there is a provision to display the results on screen. It is always advisable to keep the records of each machine independently. It becomes easy for any new person to follow-up the

TABLE 5.1
Maintenance Schedule Adopted in Conventional Carding

No.	Operation	Cycle
1.	Half setting*	Every 7–8 days
2.	Full setting*	Every 15–20 days
3.	General machine cleaning* by tenter	Twice per shift
4.	Licker-in and cylinder undercasing cleaning* by broom	Once per day
5.	Polishing of licker-in undercasing@	At the time of half setting
6.	Polishing of cylinder undercasing@	At the time of full setting
7.	Removal of dropping under licker-in#	Twice per shift for old HP card (Once for SHP card)
8.	Cylinder and doffer stripping for metallic card#	Once per day. Not for HP card
9.	General lubrication of running parts*	Every alternate day
10.	Cylinder and doffer bearings*	Oiling—every shift Greasing—at full setting
11.	Licker-in bearing lubrication*	At half setting
12.	Comb-Box—Oiling for ordinary type	Every shift. Not for HP card
	Comb Box—Greasing for ball bearing type#	Every week. Not for HP card
13.	Flat chain lubrication—with spindle oil + grease#	Every 8–10 months
14.	Grinding of cylinder* and doffer (metallic)@	Every 6–8 months (Conventional HP & SHP cards)
15.	Flat grinding (semi rigid)*	Every 6–8 months (Conventional HP & SHP cards)
16.	Licker-in wire polishing#	Every 3–4 months (Conventional HP & SHP cards)
17.	Dressing of grinding stone	After every 25–30 grindings
18.	Licker-in wire mounting	Every 2–3 years
19.	Cylinder—doffer wire mounting	Every 4–5 years—after about 1,4 lakh kg production (approx. 300,000 lb of production)
20.	Flat top mounting	Every 4–5 years
21.	Checking of card web** and waste	Every week
22.	Waste % at licker-in**	Before and after settings
23.	Nep level at card web**	Before and after setting
24.	Stop motion functioning	Once per week

@   *Only when the need arises with modern card.*
**   *Under Quality control studies.*
* and #   *No such activity required in modern card.*

history of all the machines and adjust the schedule for any preventive maintenance activities. Any maintenance operation or replacement of parts should be entered in the records of the machine in a chronological order. Even the results from the quality control, from time to time, taken before or after the maintenance schedule may also be entered to corroborate the effectiveness of maintenance operations. This helps in ascertaining the usefulness of the action taken and also becomes the guideline for future.

REFERENCES

1. Manual of Cotton Spinning – "Carding" – W.G. Berkley, J.T. Buckley, W. Miller, G.H. Jolly, G. Battersby & F. Charley
2. Elements of Cotton Spinning – Carding & Drawing – Dr. A.R. Khare, Sai Publication, Mumbai
3. Spun Yarn Technology – Eric Oxtoby U.K. Butterworth publication 1987
4. Recent Advances in Spinning Technology – International Conference BTRA
5. Forum on Tandem Carding – Textile Industry, 1962
6. Control of carding waste – J.F. Bogdan, Textile Research Journal, 1955
7. Principles of High Production Carding – G.C. Ghosh, Textile Digest
8. Tandem Card/India-rol/Crosrol Varga – MMC Bulletin
9. Neps in carding – Chr. Hohne – Textile Praxis, 1962
10. Spinning of Man-made Fibres & Blends – K.R. Salhotra, Textile Association Publication
11. Arrangement of Fibres in Card Sliver – W.E. Morton and R.J. Summers – Journal of Textile Institute, 1949, P.107
12. Fibre arrangement in cotton sliver and lap – W.E. Morton and Yen – Journal of Textile Institute, 952, Vol. 43 T-463
13. Some studies on the formation of hooks in carding – V.A. Wakankar, S.N. Bhaduri, B.R. Ramaswamy & G.C. Ghosh, Textile Research Journal, 1961, Vol. 31, P. 931

6 Features of a Modern High-Speed Card[5,8]

6.1 UNIDIRECTIONAL FEED TO LICKER-IN

Traditionally, the direction of the feed roller in card was such that the lap fleece entering the licker-in zone was more harshly treated. This was because the fleece movement was in the opposite direction to the motion of licker-in wires at the time of its first impact with the wires (Figure 6.1a and b).

The impact of licker-in moving in the opposite directions severely affected the fibres. In fact, it used to weaken them or in certain cases, even break them. In Rieter C50 cards, the unidirectional feed is employed. As shown in Figure 6.1c and d, the position of the feed plate is reversed. It is made to direct the feed the same direction as that of moving licker-in wires. This only reduces the power of impact and yet allows effective opening of the fibre tufts. As can also be seen from the bar chart (Figure 6.2), the yarn properties in terms of evenness and imperfections are improved significantly with this change.

6.2 MODIFIED LICKER-IN REGION[5,6]

In a card, it is the licker-in where the fibres are first opened. It is also the region where most of the trash in the cotton feed (chute/lap feeding) is liberated and extracted. In this case, the effective separation of trash from the lint is very important so that there is minimum lint loss at the licker-in region.

There are three important additions around the licker-in of a high production card (Figure 6.3). The deflector blade controls the amount of licker-in droppings and its composition. The carding segments assist in giving a thorough opening to the fibre tufts picked-up by the licker-in wire points from the feed. The traditional mote knives are modified in shape, size and for their setting arrangement with the licker-in. They can be set in an arc around the licker-in and the trash thus liberated is provided a direct path leading to the suction hood.

Though a fibre to fibre separation really takes place in the cylinder–flat zone, the activity of the licker-in is nonetheless very important in this context. If the licker-in is made to carry-out part of the job entrusted to the cylinder, it would ease the work that the cylinder has to do. Thus, it improves the efficiency of the cylinder and helps it in almost completing its main function—*individualization*. Hence, apart from the normal cleaning, the modified licker-in region of a modern high-speed card provides a better opening of the fibres.

One of the modifications in this region totally dispenses with the traditional set-up of the mote knives and the undercasing. In their place, there are carding segments and sharper knives. Whereas, the sharper knives assist in improving the trash extraction,

DOI: 10.1201/9780429486562-6

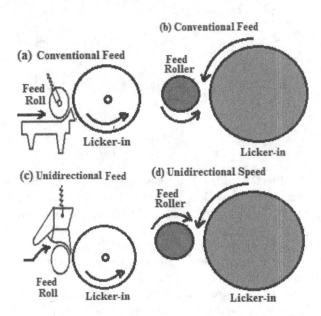

FIGURE 6.1 Feed to licker-in (conventional and modern)[5,6,8]: In conventional feed, the licker-in wires used to strike the lap fringe more harshly. In modern feed, where the direction of the lap fringe as it emerges from the feed roller and the direction of the rotation of the licker-in are the same. The action in this case is still effective but much less harsh. (a) Conventional feed. (b) Conventional feed. (c) Unidirectional feed. (d) Unidirectional speed.

the carding segments help in further improving the opening of the fibre tufts. Well opened material from the licker-in is thus properly distributed over the whole cylinder surface and this helps in further improving carding action of the cylinder. Thus, whereas the modified licker-in region improves cleaning efficiency of a card, it also assists in improving the fibre individualization later carried-out by cylinder.

In fact, an improved licker-in action helps in laying a comparatively thin layer on the cylinder. This enhances carding action. In addition to this, when cylinder speed is increased, the fibre layer on it becomes still thinner. It greatly helps in achieving higher production rates. A sharper knife is also used in between the licker-in and the cylinder (B in Figure 6.4) so that even during the transfer there is a reduction in the

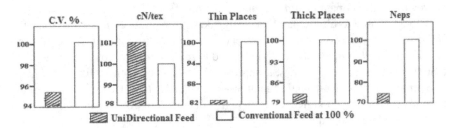

FIGURE 6.2 Benefits of unidirectional feed[5,6,8]: With unidirectional feed, the fibre damage is significantly reduced at the licker-in. This improves important yarn properties, e.g., tenacity of the yarn and overall yarn appearance.

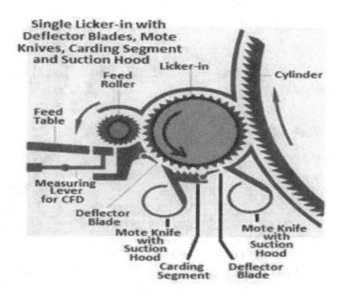

FIGURE 6.3 Features with single licker-in[5,6,8]: Modern licker-in region equipped with special features such as a deflector plate, suction hood, carding segments, etc., all of which improve the cleaning efficiency at the licker-in and, at the same time, reduce the lint loss.

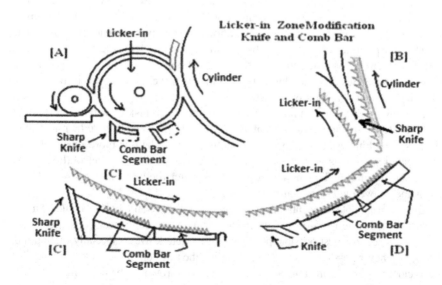

FIGURE 6.4 Comb bars with sharper knives[5,6,8]: The comb bar segments are similar to carding segments around the cylinder. The projecting saw teeth do a thorough combing job and drastically reduce the tuft size. This, together with the sharper knife, liberates a great proportion of trash from the cotton tufts.

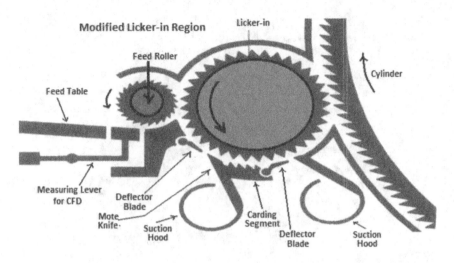

FIGURE 6.5 Deflector blade and carding segment[5,6,8]: Like carding segments around the cylinder, the carding segments are also provided in the licker-in undercasing region. With the wire points on their surface, the segments are very effective in opening the fibre tufts. This helps in significant release of trash.

tuft size. This sharp edge helps in smoothly diverting the air currents onto the cylinder region and assists in reducing the air pressure over the licker-in cover.

In another version (Figure 6.5), along with mote knife, an adjustable deflector blade is provided. This blade can be set closer to or away from licker-in wire surface. This, to a certain extent, controls the air currents around the licker. Thus, if the deflector blade is set away from the licker-in, more air is allowed to be diverted away from its surface.

The sharp mote knife placed subsequently then completes the diversion and the air thus diverted is led directly into the suction hood. It may be mentioned that the diverted air carries more trashy material. However, the diverted air, in this case, is also likely to carry a good proportion of lint.

In many modern suction hoods, sensors are provided inside the hood (Figure 6.6) and they detect the proportion of lint and trash. A light source coupled with a camera detects the whiteness of the overall collection. When there is more lint, more light is reflected and this is picked-up by the camera. It is possible to control the deflector blade setting online. This makes it possible to alter the blade setting by motorised adjustments even when the card is running.

By contrast, in old, conventional cards, though the mote knife bracket could be moved around the licker-n in a limited way, this movement was never concentric to the licker-in surface. Hence, the best possible position for the mote knives, especially the top mote knife, was never possible. As mentioned earlier, the position of the mote knife needs to be as close as possible and tangential to the material thrown by the licker-in.

In modern card, the knife, accompanied by the suction hood, can be swivelled around the licker-in concentrically. The sensor provided inside the hood continuously takes photos of the mixture of lint and trash gathered inside the suction hood and relays them to the control panel which has a set value for the proportion of lint

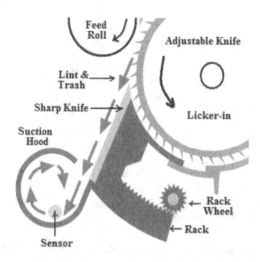

FIGURE 6.6 Adjustable mote knife[5,6,8]: As the fibres are snatched by licker-in wire points from the feed roller grip, they travel tangentially to the licker-in surface. The best position for the knife is along this tangential path.

in the extracted matter. After analyzing the signals from the sensor, the change in the setting of the knife is suitably made. This maintains the desired proportion of lint and trash and does not allow more lint in the extracted matter. In some cards, the position of knife is automatically adjusted whereas in earlier versions, manual adjustment was provided. Thus, if the lint proportion is more than a set value, the knife position is slightly made closer to the feed roller.

Other important additions around the licker-in are: a saw-tooth type of feed roller (Figure 6.5), to improve the feed grip and the feed measuring device that constantly measures the gravimetric thickness of the input fleece. The latter controls the quantity of the material entering. This, again, is very important, as apart from the variation in the quantity fed, which is likely to affect the cleaning action around licker-in, the thicker material may damage the wires.

Another modification (C60 card) is to use three licker-in rollers in place of the normal single licker-in (Figures 6.7 and 6.8). To maximize the effective transfer from one to another, their surface speeds are arranged in an increasing order from A to C. The whole set-up is available in the form of a module. Thus, the whole module can be very easily replaced by simply taking out the existing one and replacing it with a new module with fresh wires. With three licker-in rollers, the job of a single licker-in and its action at the licker-in zone are expected to be multiplied. This is because carding segments around each licker-in, together with mote knives, efficiently removes the trash particles. It is claimed that with this module, the web appears to be almost free of neps and vegetable originated trash. Under these circumstances, it is possible to work the card with higher production rates. It is claimed that a three licker-in model is best suited for open-end (OE) yarns where card production can be boosted beyond 120 kg/h.

Rieter also provides a single licker-in roller for gentler treatment. This may be of top priority when processing synthetic fibres or finer cottons, especially when

Modified Licker-in Zone

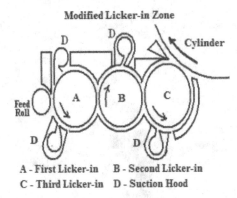

A - First Licker-in   B - Second Licker-in
C - Third Licker-in   D - Suction Hood

FIGURE 6.7   Three licker-in module.[5,6,8]

gentle fibre opening and nep reduction are more important. In addition, one licker-in roll may prove to be economical as it gives a reduced lint percentage in extracted waste. Further, fibre properties like length and strength are protected. This leads to improved yarn quality. Card production rates in this case are restricted to less than 120 kg/h.

The relative placements of the three licker-in rollers are shown in Figure 6.8. The most interesting thing is to note the relative directions of motion of the three rollers and the positioning of their teeth. At their closest distance, the wire teeth of each licker-in are pointed in the same direction, as are their directions of motion. As mentioned earlier, the surface speed for each successive licker-in roller is higher than the preceding one. The action involved here is merely of a stripping nature. Thus, the material is easily transferred from one licker-in to the next in order, and this happens simply owing to the difference in their surface speeds.

There is a distinct advantage when three licker-in rollers are accommodated in a single module. So, also, there are modules for doffer and flats. It is very easy to have a module of each type in reserve so that the module in stock can replace the one in working position in the shortest possible time. This greatly reduces the down time for the normal changeover in these regions. A special plasma coating is given to the licker-in to increase its service life. The relative wire positions of the three licker-in rollers, including the position of a suction hood, are shown in Figure 6.8.

Modified Licker-in Zone

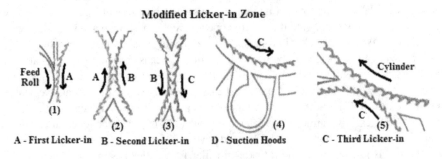

A - First Licker-in   B - Second Licker-in   D - Suction Hoods   C - Third Licker-in

FIGURE 6.8   Direction of wire points and their interaction.[5,6,8]

In another version, the licker-in placements are little different than mentioned earlier. Here the first licker-in is placed at a slightly lower height than the other two. Their relative positions and respective wire directions are shown (Figure 6.9a and b).

It can be seen that there is a progressive increase in the licker-in surface speeds. The first licker has wires with leading angle more than 90° (a concept of negative rake) whereas, the comparative relative angles for the second and third are shown in the bar charts of Figure 6.10. It can be seen that the angle from negative rake for the first licker-in is slowly and steadily increased (for second and third) in such a manner that there is more hold on the fibres. It is claimed that this arrangement gives a more gentle treatment and is quite capable in bringing about the transfer from first to last licker-in and ultimately onto the cylinder. It also carries out very effective opening and cleaning at the licker-in region. The whole treatment is still gentle because; in spite of comparatively high speed of the first licker-in, its wire teeth are pointed in slightly tilted manner (-ve rake).

This is very important, especially at the final spinning stage. If the fibres experience any harsher action at card, there is likelihood of fibre damage. This not only weakens the ultimate yarn formed at the ring frame but also deteriorates ring frame performance.

As can be seen from the bar charts (Figure 6.10a–c) all three parameters— the peripheral speed, the point density and the angle of the leading edge—on the Trutzschler DK-803 card are progressively increased so as to gradually intensify the impact of each licker-in on fibres. This improves both opening of the fibre tufts and the licker-in cleaning action.

The first licker-in is clothed (Figure 6.10d) with short needles which gently pluck the cotton tufts from the feeding system (Senso-Feed). The second and the third

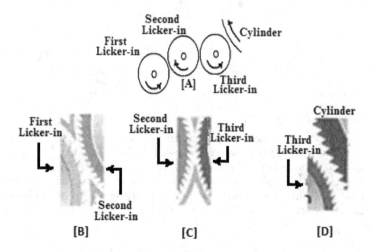

FIGURE 6.9 Three licker-in rollers[5,6,8]: Each licker-in acts on the fibre tufts. The process of reduction in tuft size is multiplied. This is accompanied by great release in trash that is sucked away. (a) Position of three licker-in rollers.[5,6,8] (b) Wires on three licker-in rollers.[5,6,8]

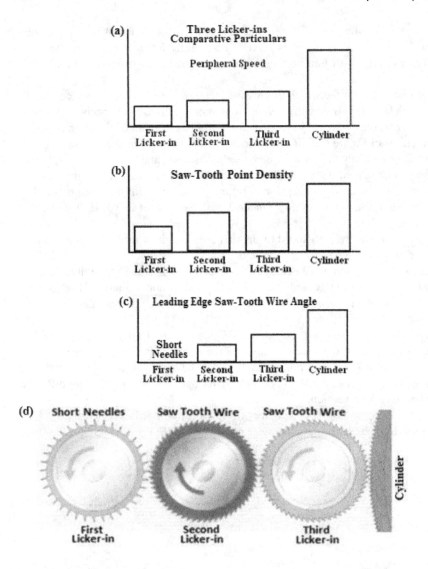

FIGURE 6.10 Relative particulars of three licker-in rollers[5,6,8]: There is a progressive increase in peripheral speeds and saw-tooth point density. Each licker-in has saw-tooth wires of a typical shape, which by their construction, become more and more intensive in their action. The size of the tooth gets progressively finer. All these, finally result in greatly reducing the mass of fibre tufts to offer better cleaning. (a) Relative particulars. (b) Relative particulars. (c) Relative particulars. (d) Three licker-in rollers.

licker-in rollers have saw-tooth wires on their surfaces. By the time the material is about to be transferred onto cylinder, it is almost in the form of a thin web. It may also be pointed out here that ultimately the action of the third licker-in is as strong as that with a single licker-in.

By keeping the action of the first licker-in slightly on the milder side, the first impact that the fibres receive, when released by the feed roller, is made comparatively milder. It may be mentioned here that though each succeeding licker-in acts more strongly, the act of fibres switching-over from one licker-in to the next is more of a transfer type. The succeeding licker-in rollers, however, reduce the fibre film in progression. The last or the third licker-in finally makes this fibre film quite thin and presents it to the cylinder in a much opened form. This helps in improving the cylinder carding action. Thus, it may be summarized that the use of three licker-in rolls not only improves the cleaning in the licker-in zone but also reduces the burden on the cylinder. This facilitates easier cylinder carding action. In addition, with the three licker-in rollers, the three suction hoods provide additional cleaning potential for removing the trash.

Lastly, a well-opened fibre material entering the cylinder zone helps in very effective nep reduction and improves the card web appearance.

6.3   CARDING SEGMENTS[5,6]

The overall positioning of the various carding segments and suction hoods is shown in Figure 6.11. There are carding segments placed before the main carding action and in between cylinder and flats, called *pre-carding segments*. There are also additional carding segments placed in between cylinder and doffer. They are called *post-carding segments*.

As can be seen, the segments are stationary and obviously require replacement when the fine saw-tooth wires on them wear out. The wire-covered surface of the

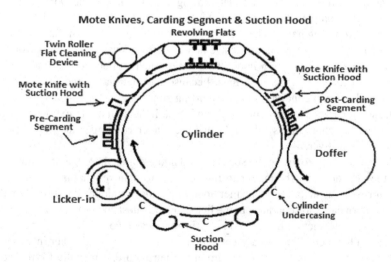

FIGURE 6.11 Positioning of carding segments and suction hoods[6,8]: Though the main carding action takes place between cylinder and flats, the very useful additional carding surface is provided by carding segments. As each segment is also accompanied with a suction hood, along with continued carding action, the impurities released in this process are immediately sucked away.

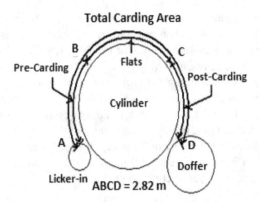

FIGURE 6.12   More area available[6]: The wire covering surface due to carding segment is increased. This provides larger area for carding action.

carding segments provides an additional carding power to what cylinder and flat can do together. This increases the power of card to take-up a higher rate of input. This, in turn, helps in achieving higher production rates. A much larger carding surface area is provided (Figure 6.12) by the use of carding segments. The total peripheral distance from point A to D is around 2.82 m. Addition of carding segments gives higher pre-& post cylinder-opening and this helps in giving more intense carding.

This facilitates increased production rates. An enlarged post carding area also ensures cleaner sliver and improves fibre individualization. The carding segments, especially the pre-segments, actually provide the additional carding operation as a preparation for the main cylinder–flat carding action. Further, the work would have been only half complete had it not been for the suction hoods positioned immediately after the carding segments. The loosened-out impurities and trash thus released by the action of these segments are conveniently carried away through controlled air suction created at the suction hood. This greatly helps in easing the main carding action between the cylinder and flats, because the material is received in a more opened and cleaned condition in main carding zone.

The suction hoods are also provided underneath the cylinder to extract the lighter impurities and short fibres released during cylinder running, thus improving the quality of card sliver.

The purpose of providing the suction hood at the point where the flats move away from the cylinder after the main carding action is to remove what is left on the flats in the form of flat strip. It may be mentioned here that a unique provision is made in latest modern cards to separate-out the licker-in and flat strip waste by carrying them through separate suction hoods and later through waste conveying pipes to suitable respective chambers. This, to a great extent, helps; because the licker-in waste and flat strip waste are basically very different in their nature, especially for their composition, the former being very dirty in nature.

Though the post-carding segments do not directly contribute to the main carding action; they almost finish the job of completing the opening and cleaning of the material coming from cylinder-flat zone. Thus, they are responsible for ensuring the deposition of a really clean, fine and filmy web on doffer. This can be seen from

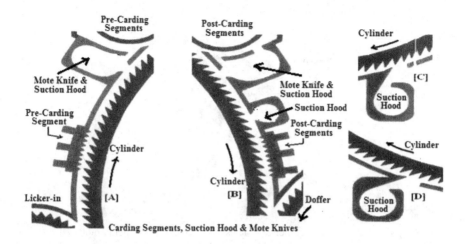

Carding Segments, Suction Hood & Mote Knives

FIGURE 6.13 Carding segments and suction hoods[6,8]: These two are important additional attachments around a modern card. The carding segments increase the power of opening of fibre tufts to reach very close to a state of fibre-to-fibre separation.

the quality of the web coming from doffer on any high production card. It may be said that, with carding segments, the web finally delivered by the doffer is almost transparent, nep-free and trash-free.

Figure 6.13 shows the wire teeth positions and relative placement of the carding segments more elaborately. Part A indicates the pre-carding segments and suction hood accompanied by a mote knife. The purpose of the mote knife here is more to bifurcate the lint and trash and allow the latter to be conveniently pulled away by the controlled suction through the hood.

Part B in Figure 6.13 shows the position of post-carding segments and two suction hoods; one of the hoods is closer to the outgoing flats. Here the mote knife plays a similar role.

The other suction just prior to these segments is merely to take full advantage of removing any remaining loosened-out trash in the material before it is carried on to the doffer.

Parts C and D are the suction hoods placed under the cylinder. These suction hoods are also provided with bifurcating sharp edges, though not in the form of mote knives.

Both the positioning and the setting of these knives with the cylinder are very important.

In fact, the closer setting of knives gives elevated carding power, dislodges some of the impurities in the web and leads them into the suction hoods placed immediately after. The impurities are then carried away into waste chambers. Thus, while the carding segments improve the carding action, the knives improve cleaning.

A special feature of Rieter's C50 card is the Trex system. As shown in Figure 6.14, these are pre-carding segments. The wire-covered surface of these segments, while opening the fibre tufts, helps in very effective removal of short fibres and dust.

As mentioned earlier, every carding segment has to work in conjunction with a suction device. Thus, when the short fibres and the dust are liberated owing to

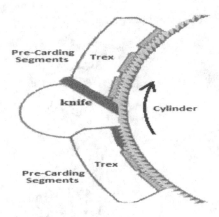

FIGURE 6.14   Trex-carding segments[5,8]: These are typical attachments mounted over the cylinder in carding in improving the carding action.

the action of the carding segment, they are not allowed to recombine. Advantage is immediately taken of this separation and they are sucked away. The Trex system (Figure 6.15) of the carding segment is very successful in controlling thin and thick places in the ultimate yarn formed. It is claimed that, with their use, there is substantial reduction in neps. This is because these carding segments are able to assist with a thorough opening of the fibre mass around the cylinder.

Another interesting feature of Rieter's C50 card is incorporation of the Integrated Grinding System (IGS). The system provides automatic grinding of the cylinder during normal working time. It is claimed that with the use of new-generation wires used on the cylinder, no spark is produced during grinding, thus avoiding any damage due to fire.

The IGS system is expected to maintain or even improve the web quality at higher production rates. As regards the Trex, an improved carding helps in better transfer of fibres onto the doffer. The effective opening action substantially reduces major yarn defects such as thick–thin places and neps in the card sliver. The ultimate yarn formed is thus much superior.

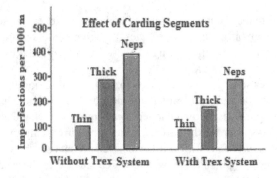

FIGURE 6.15   Effectiveness of Trex system[6,8]: It has been proved that with such system, the yarn appearance is positively improved through reduction in yarn imperfections.

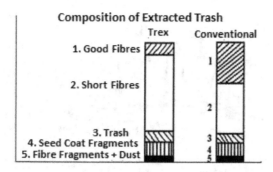

FIGURE 6.16   Performance of Trex[6,8]: The system is more economical in two respects. Firstly, there is a sizable saving in the long fibres, and secondly, the short fibre content in the trash is markedly higher.

It can be seen from the bar chart (Figure 6.16), the use of the Trex system results in a significant reduction in the yarn imperfections. As claimed by the manufacturers, there is a reduction of up to 40% in total imperfections. When the power of the card is increased, it is possible to speed-up production rates without affecting the quality of the card sliver. Rieter's *Vario-set* blow room, together with the C50 card, achieves the required quality of sliver and hence that of yarn. The Trex system ensures very selective extraction of short fibres, trash particles, seed-coat fragments and dust. As compared to this, the good fibre loss is comparatively much less. The percentage of short fibres in extracted waste, in particular, clearly shows that the fibres extracted into the suction hood are less than 12.5 mm. In the conventional system, there is comparatively higher loss of good fibres as well as lower extraction of short fibres.

6.4   FLATS AND THEIR DRIVING[5,6]

The flats on a DK 760 card are driven by a toothed belt (Figure 6.17) and are pressed against the flexible bend. The belt ensures smooth, secure and snatch-free running of the flats over the flexible bend. The operations are further eased

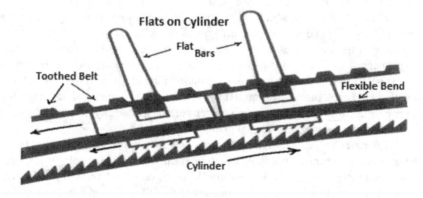

FIGURE 6.17   Flats and their driving[6,8]: The use of a toothed belt ensures a smooth and jerk-less running of the flats. It also helps in easy dismantling and refitting of the flats.

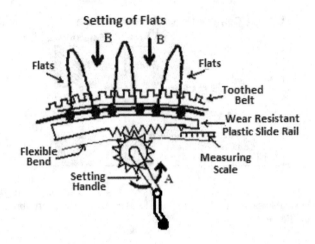

FIGURE 6.18   Flat setting[6.8]: A setting handle accompanied with plastic slide rail facilitates the setting of flats with the cylinder. The provision of a measuring scale helps in reaching the required accuracy.

by facilitating replacement of the flats without any tools. Thus, whenever a flat is required to be replaced, it can be simply taken out from the belt and a new flat can be inserted in its position. In Trumac (Trutzschler), a mechanism is provided to precisely set the flats.

The setting handle can be turned in direction A (Figure 6.18) so as to lower the flats (B) or vice-versa. This is achieved by providing a special wear resistant toothed plastic slide which is made to rest on a flexible bend. The slide works with another toothed wheel to which this setting handle is attached. The movement of the handle makes the plastic slide to either raise or lower the flats. A measuring scale is provided to bring about the exact change so as to precisely set the flats.

Like cylinder or doffer, the flats also can be reground on the machine. This is made possible by special mountings which can be fitted to the card to hold the grinding device. In this case, a guiding unit ensures that the flats are accurately presented along the surface of the grinding roller. The adjustments for flat settings are reproducible and these can be recalled for any type of flat wire.

Wearing of the card wires is inevitable and, especially in high production carding it is faster. Rieter's *Integrated Grinding System* is a unique feature of their cards (C50 series). It ensures that the flat wires are maintained sharp at all times. It is found that with this system, there is substantial nep reduction and removal of trash, and this ensures a uniform sliver quality over a longer period. It is a fully automated system that eliminates any down time for flat grinding operation.

In the Trutzschler TC 5 card, magno-tops have been introduced (Figure 6.19). In place of classic aluminium bar flats, which hold the tops by means of clips, special *neodymium* magnets are used. As mentioned earlier, on a modern card the flats can be easily dismantled and replaced without any tools. While fixing, they are required to be merely pressed into a toothed belt. In this manner, they are quite firmly. But still, the magnets used provide a firm and secure hold on the tops.

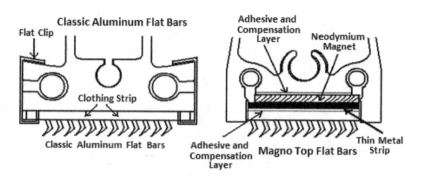

FIGURE 6.19 Aluminium and neodymium flat tops[6,8]: In place of classic aluminium flats, neodymium flats are used in some modern cards. With them, a special type of magnet holds the flat in place of conventional clips. This makes it easy to dismantle and refit the flats without any tools.

6.4.1 REVERSAL OF FLATS[7,8]

In a conventional card, when the flats make their entry from the licker-in side, they are all fresh and clean. But as soon as they enter the carding region, they start receiving the material from the cylinder for immediate carding action. As such, in a conventional card, the material delivered from the licker-in is still in the form of small tufts. Therefore, as soon as the flats start taking part in the carding action, they are flooded with fibres. On an average, it is found[2,3] that by the time a flat occupies the fifth or sixth position from the licker-in entry point, it become significantly filled with fibres and trash. In addition, there is quite a large proportion of dirt released when the cylinder carrying small tufts of cotton meets the flats in the beginning. The released dirt also gets into the flat clothing. Assuming that there are about 42 flats around the working surface of the cylinder, the journey of the (say) sixth flat to its 42nd position does not happen to be much fruitful in extracting further trash. Especially the fine dirt which the sixth flat might hold at that point will have to be carried further till the same flat occupies the position beyond 42nd flat. This is because only then the embedded fibres, trash and dirt can be taken away from the flat surface so as to freshen and clean them again.

When the flats are reversed, they enter the carding zone from doffer side. Obviously, they are quite fresh and clean at the time when the material on the cylinder is about to leave the carding zone. The clean and fresh flats entering from doffer side, thus provide more opportunities for final cleaning of the fibre film before it leaves the carding zone. As the flats run backwards (are reversed), they again get loaded with fibres and trash. When they get closer to the licker-in, they are almost fully loaded with fibres and trash. As mentioned earlier, it is at this point that the tufts received from the licker-in by the cylinder start facing the carding action. Owing to the suddenness of the action, a lot of fine dirt is released at this juncture.

Though sufficiently loaded, the flats are still capable of catching this released dirt (Figure 6.20a). However, they don't have to carry it over a longer distance. This is because, immediately thereafter, they curve around and leave the carding zone from

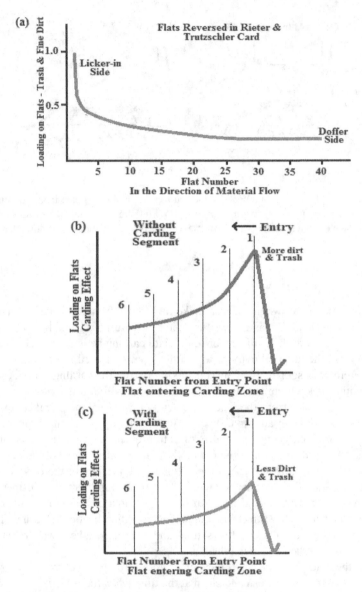

FIGURE 6.20 (a) Dirt carried by flats[5,6,8]: When flats are reversed (they enter from doffer side), they are fresh and do the finishing job of carding the material far better. The fibres carried by the cylinder thereafter are cleaner and better carded. (b and c): Loading on flats (with and without carding segments)[5,6,8]: The contribution of the carding segments to the work of carding is quite significant. They assist the main carding action between the cylinder and flats. Along with carrying out carding action, the released dirt and impurities are immediately sucked away. This leads to less accumulation of dirt and trash in the flats. (b) Without carding segments and (c) with carding segments.

the licker-in side. They are quickly cleaned by the flat cleaning unit and become fresh and clean. The released dirt thus gets full opportunity to be removed.

In a high-production card, there are pre-carding segments (Figure 6.20b and c) which greatly assist the opening of the fibre tuft. The suction provided around such carding segments also improves the cleaning. This very effectively removes fine trash and dirt. A comparatively clean material is carried by the cylinder as it enters the carding zone (from the licker-in side). This is only because of the contribution of the carding segments[2,3] and the suction hood around it. As a result of the use of the carding segments in the pre-carding zone, therefore, the release of the dirt before the cylinder-flat carding zone is significantly reduced. Whatever small percentage of dirt that still remains is easily captured by the reversing flats as they leave the carding zone from licker-in side. In general, more opening of the tufts at the carding segment and uniform distribution of the material over the cylinder surface reduce the loading on the flats (Figure 6.20b and c). The studies carried out in this regard confirmed not only that the trash content in the flat strip was considerably reduced, but also that the good fibres lost in the form of flat strip were comparatively less.

6.4.2   FLAT CLEANING SYSTEM[5,6]

On a modern card, a provision is made to collect and remove the flat strip and licker-in droppings separately. Thus, the dirtier droppings around and under the licker-in and carding segments are separated from the comparatively cleaner flat strip and this allows cleaner waste to be reused for waste trade.

On a DK 760 card, a twin roller system (Figure 6.21) is provided for cleaning flats. These are self-cleaning types of rollers and provide gentle cleaning of the flats' surfaces. After the flats leave the carding zone from the licker-in side, the flat tops are exposed to a mild and yet effective stripping action of the revolving stripping brush, which lifts-off the strip and embedded trash. The stripping roll, in turn, is cleaned by a fast-revolving brush with the help of sharp knife edge (K). The matter that is stripped-off is ultimately guided on to a suction duct, where powerful suction carries away the fibrous matter along with the trash. Thereafter, the flats return to their entry point from the doffer side to carry out the fresh carding job.

6.5   WEB DOFFING[5,6]

In any high production carding, the method of web doffing is always of high importance. The web coming from the doffer is thin, transparent and delicate, and therefore has to be handled carefully during doffing. It is necessary to see that the web produced should be a neat, clean and free-form sliver without breaks. The edges of the web, especially, have to be handled with a lot of caution. If there are any contaminants in the web, they are very likely to cause disturbances during sliver formation. Trutzschlers, in their DK 760 card, have designed a special doffing (Figure 6.22a and b) with a web guidance system. A doffing roll system automatically guides the web from doffer to *web speed*. Initially, the web is taken-off the doffer by a stripping roll and is then passed on to the web speed. This helps is avoiding any extra piecing aids. The *web speed sliver-former* condenses the web and guides it to a measuring funnel.

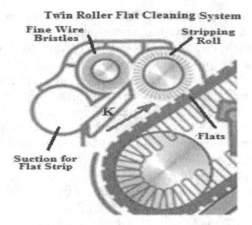

FIGURE 6.21 Flat cleaning system[6,8]: The twin roller provision doubly assures that the flats as they come out from the licker-in side are fully cleaned. This increases their effectiveness when they come back again for their carding action from the doffer side.

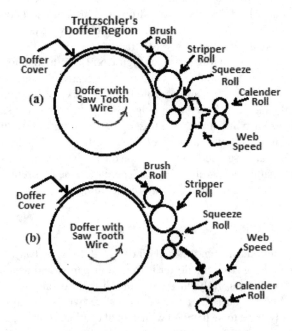

FIGURE 6.22 Web doffing: As the web coming from doffer is very thin and delicate, the web doffing at a high speed is a very skilful operation. The web speed in DK 760 card manages this very efficiently. (a) 'Web Speed' in normal position and (b) 'Web Speed' during piecing.

For mending sliver break or guiding the web to calender rollers, or even for inspection of web, the web speed can be tilted in its position (Figure 6.22b).

It may be noted here that the typical individualization of the fibres achieved on the cylinder is disturbed during the transfer of the fibres from cylinder onto the doffer. Further, the required parallelization is almost achieved when the fibres are held by the cylinder wires. When they are condensed on the doffer, this disappears. However, a random orientation is very much necessary during this deposition on the doffer so as to form a web strong enough and ready for doffing. Another undesirable effect which occurs is the bending (hooking) of fibres during the transfer. Both the whipping action of the cylinder and the fibre elasticity are responsible for this.

It is well known that, on an average, the fibres move around the cylinder 1215 times before being transferred. Apart from providing additional carding, this intermingling offers the extra advantage of allowing fibre-to-fibre blending.

6.5.1   INNOVATION IN WEB DOFFING ON RIETER CARD[5]

A novel idea is used by Rieter in web doffing. The take-off roll (Figure 6.23b) is set very close to the doffer and is covered with special type of wires, very much unlike conventional saw-tooth wire. While the teeth are quite efficient in peeling the web off the doffer surface, they do not hold the web. This is why it is possible for the delivery rolls to easily get hold of the web peeled-off by the take-off roll. The novelty of this whole operation is the subsequent collection of the semi-transparent, thin and filmy web by the two aprons. In the direction shown (Figure 6.23a), the web across the whole width of the delivery rolls is collected by these aprons, partially condensed and centrally delivered in a somewhat rope-like form.

This serves two important purposes. Firstly, the early consolidation of web prevents its falling down in the region between the peel-off point and the final disc rollers (calender rollers in a conventional card). Secondly, and more importantly, this consolidation prevents any undue stretching. Thus, it prevents any possible medium- and long-term irregularity. In fact, the disc rollers, while allowing the sliver to pass through, also control the regularity of the out-going sliver.

6.5.2   INNOVATIVE MAGNETIC WEB CRUSHING[5,6]

When the conventional crush rolls are used, they must either be offset or they must allow their shape to be distorted when heavy impurities pass through them. Traditionally (Figure 6.24b), the pressure for crushing the impurities is applied at the two sides of the shaft, which is held by the bearings on either side. Therefore, when any heavy impurity passes through, the rollers bend slightly.

This bending, in turn, affects the pressure on the entire length of the crush rolls. Thus, the pressure exerted by the top crush roller on bottom becomes uneven across its whole length. The magnetic web-roll gives uniform and even pressure along the entire width of the card web (Figure 6.24a). A permanent core is designed into the 3-inch top roller.

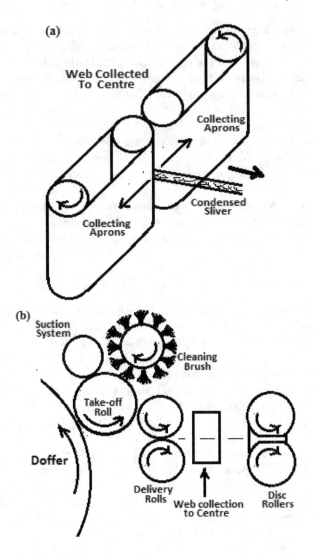

FIGURE 6.23   Central web collection (Rieter)[5.8]: In any high production card, the web from a fast-revolving doffer has to be redirected or lifted off the surface by using a *lift-off* or *take-off* roll. In addition to this, in Rieter, there is a supplementary central web collection system using rotating aprons. This prevents any web distortion or stretching. (a) Collecting aprons (Rieter) and (b) web doffing (Rieter).

The magnetic forces are, therefore, distributed evenly across the length of the roll. Even when web conditions vary, the force of attraction between the top magnetic roll and bottom doffer take-off roll remain the same. The uniform distribution of this pressure eliminates any inequality of load on the bearings and their life is improved. Thus, uniform pressure is responsible for breaking and crushing trash particles of any size. Subsequently, the powdered trash tends to get separated before formation of a sliver in the card.

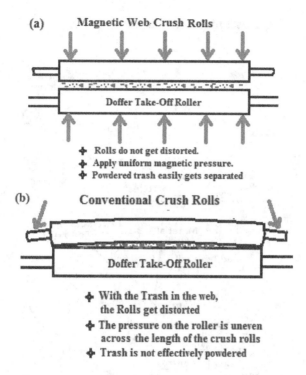

(a)   Magnetic Web Crush Rolls

Doffer Take-Off Roller

✤ Rolls do not get distorted.
✤ Apply uniform magnetic pressure.
✤ Powdered trash easily gets separated

(b)   Conventional Crush Rolls

Doffer Take-Off Roller

✤ With the Trash in the web,
  the Rolls get distorted
✤ The pressure on the roller is uneven
  across the length of the crush rolls
✤ Trash is not effectively powdered

FIGURE 6.24   Web Crushing[5,6]: In any high production carding machine web crushing is very important function. The crushing is accomplished by putting pressure on the crush rolls. The impurities are crushed and powderised. This means that the fine dirt granules can easily be dusted off. (a) Magnetic web crushing. (b) Conventional web crushing.

6.6   SLIVER REGULATION[6]

For optimal performance of any card, there must be a controlled and constant quantity of material entering from the feed side. This basically has two objectives. Firstly, it is necessary to see that the card gives the best performance in terms of both cleaning and individualization of fibres. A controlled feed (Figure 6.25a) ensures that all the important organs, such as the licker-in, cylinder, etc., receive a constant quantity of material to work on, thus stabilizing their corresponding actions. The second important role that the constant feed plays is to ensure that the outgoing sliver measures uniformly along its length. This helps in maintaining the final yarn uniformity.

To meet the second objective, the sensing device is actually the sliver condensing (measuring) funnel situated at the delivery end. Depending on the thickness variation in the sliver, the variation is sensed by the measuring funnel. It is then conveyed through the controller to the feed roll drive.

The drive to the feed roll is ultimately altered in proportion to suit the corresponding variation sensed. Thus, for heavier than normal weight of the sliver, the feed roller speed is reduced. However, it must be remembered that the variation is sensed at the delivery side. It is conveyed electronically and therefore very fast. Still, the correction is done for a totally different portion of the material. In comparison to the

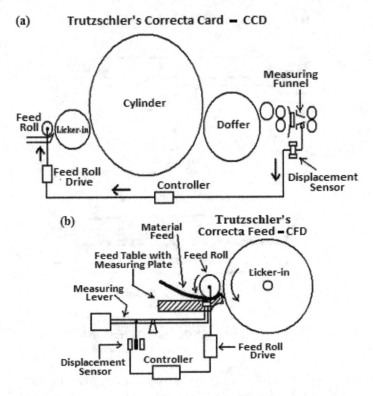

FIGURE 6.25 (a) Sliver regulation[6,8]: This device works on the principle of auto-levelling the out-put mass of the sliver. The measuring funnel, situated at the delivery funnel, measures the variation in the mass per unit length and the signals are conveyed to the feed-roller speed. (b) Controlled feed at licker-in[6,8]: Whereas the sensing of unevenness of the lap is done on the feed table, the signals are accordingly conveyed to vary the speed of feed roller.

above, the device (CFD) controlling the feed to the licker-in (Figure 6.25b) has more direct impact. Here, the material is sensed as it passes under the feed roller. Any variation in the thickness of the material being fed is immediately sensed. The signals are relayed electronically at considerable speed to bring about the corresponding change in the speed of feed roll.

6.7   LENGTH VARIATION CURVE[6]

It is seen from the graph (Figure 6.26) that, the effect of the Correcta Card (CCD) and Correcta Feed (CFD) on controlling the card sliver variation is quite significant. The CV% of card sliver for short, medium and long cut length has been significantly reduced. The graph also reveals that CFD definitely has an edge over CCD. When they are used together, there is substantial reduction in short and medium term variation. The differences among the curves—one without regulation, the second with only CCD, and the third with both CCD and CFD—are slightly reduced when long-term variations are considered. This means that both CCD and CFD, when used

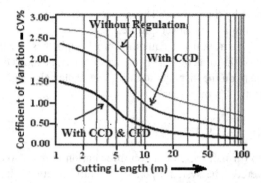

FIGURE 6.26   Length variation curve[6,8]: Without automatic sliver regulation, the coefficient of variation is very high. When both CCD and CFD are used, the variation can be substantially controlled.

together, are more efficient for short and medium term variation. The yarn produced, in this case, is more uniform and regular.

6.8   NEP CONTROL IN THE HIGH PRODUCTION (H.P.) CARD[6]

On the Trutzschler TC5 card, provision is made to constantly monitor the web quality. This avoids elaborate testing. The sensor (Figure 6.27a) is actually a digital camera placed just below the take-off roll on which the fibres transferred from doffer are in a very thin layer form. The camera is capable of taking pictures at a speed of 20 shots per second. The instrument works (Figure 6.27b) on the principle of reflected

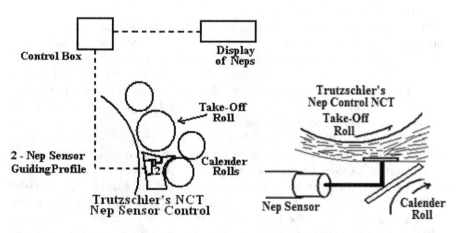

FIGURE 6.27   Nep measurements[6,8]: The presence of neps in the card web is one of the barometers for deciding the quality performance of a card, so the sensing of neps and their detection becomes of vital importance. The sensors are placed very close to the surface of the take-off roller and continuously detect and count the neps. (a) Neps measurement and display and (b) Neps measurement and display.

light which is caught by the camera. The camera itself moves along the whole width and is encompassed in a fully closed profile.

Viewing the neps through the camera almost duplicates seeing them with the naked eye. It is possible to record the neps, trash particles and seed coat fragments, as well as to see them on the display monitor. The display also shows the distribution profile of neps, trash, etc., on screen.

6.9   CARD INFORMATION SYSTEM[6]

This is a system to furnish the production information in order to assure the quality of the sliver delivered. The system operates in both directions. It monitors a group of cards. If any card goes out of preset tolerance, that card is immediately stopped. This is done parallel to its normal function of evaluating its performance in terms of production and quality. All Trutzschler cards—DK 740, DK 760 and DK 803—are prepared for their connections to the card information system *KIT* system (computer controlled production and quality-data collection system). What is then required is an additional interface. It is also possible to have a host system [Central Information Trutzschler (*CIT*)].

The data is displayed in the form of graphics or tables on a colour monitor. Even the measuring units, such as metres or yards, kilograms or pounds, ktex, count in Nm, or Ne can be selected. There is no need for any separate programmer as the programme itself is user-friendly. The guidance is given through a dialogue box. The KIT system is equipped with a special computer, colour monitor and printer with corresponding software package and communication network. Special cables up to 600 m are supplied to reach the vicinity of the cards. Through KIT, all the important data related to operational status, production and quality of card sliver are displayed continuously, and the colour graphics and tables show the instantaneous situation or an integrated value over a time period.

It is possible to have a spectrogram for the card faults and each spectrogram is stored every hour. The data from up to 72 spectrograms can be stored and thus can be made available for inspection at a later stage. When there is a deviation from pre-determined values, an alarm signal is given. If there are defects in the card sliver, it is possible to get the information to assess the possible causes. Based on the length variation curve, the information is made available on the regulation system which is responsible for controlling this variation.

The production values are displayed for each card on a monitor and the current status for a group of cards is shown. The production data is automatically summarized over a shift or day, over lots or based on quality. If the efficiency for any card falls short, a comprehensive analysis is made available. All mal-functions are recorded and listed for the time duration in chronological order for suitable counter-measures.

On the other hand, on the Rieter C50 card, there is a complete system called ABC and C-control for both blow room and card. It monitors and controls all the machines in a sequence as well as the pneumatic fibre transportation. The system has programmable logic control (PLC) for real-time processing and is interfaced with a computer. The LAN is used for data exchange. The C-control used on the card is a subsystem. The colour monitors provide the touch screens and make the system user-friendly. All functions can be directly accessed by choosing appropriate function buttons.

The data related to production, quality, alarms, shifts and maintenance can be recorded and displayed. The data is made available in both numerical and also in graphical form, so the reports can be generated and printed instantly, either as graphs or in a list form. The data can also be stored on a hard disk for future reference for processing off-line. It is possible to identify a group of cards processing the same material and the changes in technical or production particulars can be made for the whole group. A further extension of this is to control the operations of a complete plant. Rieter provides such networking on request. This data in a substantially condensed form is then transferred via standard file transfer protocol to a server computer.

6.10   CAN CHANGING AND AUTOMATIC CAN TRANSPORT[6]

In Trutzschler card equipment, it is possible to convert any high-speed coiler-canto an automatic high-speed can-changer unit. The can sizes can also be varied. The system, known as KHC, automatically transports the full and empty cans. The operation of feeding empty cans and removing full cans is done from the front so as facilitate ease of operation. There is an automatic sliver-cutting device which operates during the can changing, and a precise overhang length is maintained. This length is automatically guided into the next empty can. The KHC system is adaptable to the can height and a provision is made for a separate servo motor.

The TC-C can changer on Trutzschler Card TC 11 is specially designed for a can size of 1500 mm (height) × 1000 mm (dia.) and is most suitable for high-speed delivery. It stands independently and separately in front of the Trutzschler card. A small service space is provided between a card and can changer and it considerably reduces the approach distance from one card to the next. A separate drive is given to the device to allow fine draft adjustment between card and can changer. This helps in giving absolute control over the tension drafts to avoid sliver breaks in this region. During each can change, there is an automatic sliver separation. Thus, a fresh sliver-end enters the new empty can at every can change.

Depending on the need and the floor space available, a special transport system is made available. For single lot processing, the designing of such a system is easier. In this case, the fixed rail-tracks are used to transport the cans from card to draw frame. However, if multiple lots are running on a set of cards or if transporting distances are longer, Trutzschler has developed the *Canny One* system, which is more flexible. It uses several transporters, each carrying one can. The data related to the transport path is transmitted through infra-red signals. A *user dialogue* is provided to allow the change of lots. It is possible to avoid any manual intervention. There is also a provision to remove sliver remains from the empty cans.

6.11   EXACTING CARD SETTINGS[6]

The distances of all carding elements from the cylinder are very important and involve crucial settings on any high production card. This is because they significantly influence the card sliver quality. On a conventional card, all the card settings were made when the machine was stopped. They, in a true sense, did not take into account the dynamic changes in terms of centrifugal force or even the change in the temperature of working organs. It is observed that these factors do affect the card settings during processing.

The Trutzschler TC 11 card incorporates setting optimizer T-Con, which is supposed to control five different functions: (1) displaying the most important process variables controlling carding quality, (2) maximizing safety against clothing contact, (3) optimizing sliver quality making correct settings, (4) analysing operating status as a pre-requisite for controlling settings and (5) using spacers to enable fast reproduction of accurate settings.

The optimization tool T-Con is a software package integrated into card control. It has special sensors to measure values for parameters such as current temperature, speed of carding elements. etc., and it also carries-out the analysis. On the screen, the current parameters along with those best to optimize are displayed for the type of fibres being processed. Thus, the flat distance can be changed during processing within the shortest possible time. Coloured spacers are provided for changing the settings of carding elements.

If the settings are too narrow, a warning appears on the screen. If there is even a very small contact between the metallic wires, signals are relayed to immediately stop the card. Special sensors are provided to constantly monitor the temperature of the cylinder and any undue rise is immediately displayed in the form of a warning.

6.12   INDIVIDUAL MOTOR DRIVE[6]

The concept of individual drive is not unknown to the textile industry. There are several drawbacks to group drives e.g., huge belts running from the overhead pulleys to the machines on the ground, pose a continuous risk to the workers in the department. There is a great power loss due to belt slippage. Serious accidents are likely to occur when any of the belt breaks. Equally important is that a proper lighting system on the working parts of the machine cannot be planned.

These are replaced by individual drives. But so far the concept has been mostly restricted to using an individual motor for each machine. In high production carding, it becomes essential that the speeds of different organs be controlled very precisely. This is because it is very important to maintain the relative speed differences between certain organs like the licker-in, cylinder and flats. Equally important is to give controlled speed drives to certain organs like chute feeding mechanism, the feed roller doffer and related doffing assembly, the calender rollers and coiler calendaring, etc.

The second aspect, in particular, becomes an important part of any high-production carding where controlled feed allows a far better quality of card sliver in terms of both its cleanliness and regularity.

Therefore, providing individual motor drives to many of these elements has become essential. In the Trutzschler DK 803 card (Figure 6.28), various drive-motors and transmission elements are incorporated. Thus, special AC motor drives are given to the cylinder, licker-in and flats. In such transmissions flat belts are used.

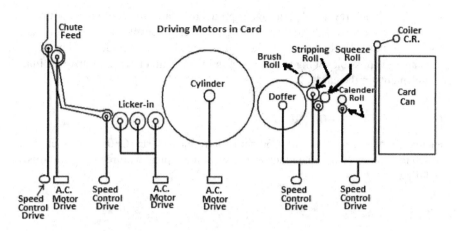

FIGURE 6.28   Individual driving motors[6,8]: Providing individual motors has two distinct advantages. Firstly, the speeds of the individual organs can be independently controlled and changed when required. Secondly, the power requirements are changed and this reduces heat generation.

This is because, at the start, there is always some belt slippage. The flat belts are specially made so that they require little maintenance and give long life.

The web doffing is equipped with a separate variable drive. This enables the card to operate with ideal draft at any speed. Further, this draft adjusts itself automatically during start-up and slow-down. This greatly helps in maintaining sliver regularity throughout the production. The motors themselves have no carbon brushes and hence are almost maintenance-free.

These motors have excellent dynamic properties and hence the speed characteristics are unaffected by the load. This improves the short-term sliver regularity. When there is an option to choose between the can coiler and can changer, the system can be provided with a servo drive. The graph in Figure 6.29 shows how the power is reduced relative to sliver production in kilograms per hour. It is well known that when the production rate is increased, it becomes essential to increase the cylinder

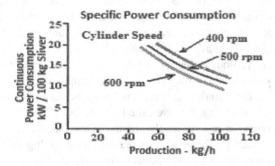

FIGURE 6.29   Power consumption[5,6,8]: With increase in production rate, especially at higher cylinder speed, there is reduction in specific power consumption.

speed. The graph reveals that at higher production rates the continuous power consumption in terms of kilowatt per 100 kg of sliver processed seems to be reduced, and it is found to be the lowest for the highest speed of the cylinder used in the trial. This again is possible owing to the use of individual motors used to drive all important elements in the card.

6.13   RIETER'S C60 AND C70 CARDS[5]

Rieter, after their success with C50 cards, introduced series C60 and C70 cards, for which they received a tremendous response. Some typical features of these cards are as follows.

6.13.1   C60 CARD

A high performance C60 card, while maintaining some earlier concepts, incorporates some new concepts—a higher working width, an increased carding angle for wires, smaller cylinder diameter, introduction of integrated grinding system (IGS) and module system for helping easy maintenance. With all these, it is possible to achieve higher productivity through higher speeds and lower downtime. They all help in increasing trash extraction, improving sliver quality and maintaining its consistency. It is thus possible to attain a production rate up to 180 kg/h (about 50% higher than the C50 card).

The C60 card can be equipped with a single licker-in or triple licker-in system and they are interchangeable. Thus, the whole module can be replaced or removed. The modular system helps in quite appreciably cutting down the time to change from single to triple licker-in. Another very interesting feature of C60 technology is the combination of card and draw frame. There are two versions possible: C60-SB where there is no auto-levelling, and C60-RSB which provides for auto-levelling. This is really a novel idea as the technology of Rieter's draw frame has been successfully put to use for a single sliver drafting. In the C60-RSB combination, special attention has been paid to the can-changing mechanism. The normal practice on standard draw frame is that the speeds are reduced to absolute minimum (10 m/min) during can change and this results in deterioration in sliver quality. As against this, in the C60-RSB combination, the can changing takes place at a much higher speed of 100 m/min. Simultaneously, the *levelling actuation point*, which is a function of delivery speed, is automatically adjusted. This guarantees a high level of uniformity at all times.

Transfer factor (TF) quantifies the proportion of fibre mass transferred onto the doffer with every revolution of the cylinder. For example, with a 100% transfer factor, all the fibres on the cylinder get transferred onto the doffer when they come close to it. For normal functioning, however, the transfer factor ranges from 5% to 15%. Basically, the production rate markedly influences the transfer factor. As the output rate increases, the transfer factor also increases. However, higher fibre load on the cylinder, in this case, has an adverse impact on the sliver quality.

As compared to C50 card [7.7 rev. – TF = 13%], the fibres on the C60 card remain for longer time [10 rev. – TF = 10%]. Thus, in spite of the smaller diameter of cylinder, the carding intensity remains almost the same. As claimed by the manufacturers, it is possible to achieve different production rates ranging from 180 kg/h for denim and 60 mm polypropylene, around 140 kg/h, for knitting yarns and polyester/cotton blends, and around 130 kg/h for regenerated cellulose.

The cylinder has a quite high surface speed of 40 m/s. Its distance, on average, from the parts around it is just 0.1 mm. With such closeness, it is absolutely necessary that the whole cylinder is maintained perfectly concentric. Even at a very high speed of 1000 rpm, the concentricity error with the C60 card is substantially low. The cylinder bulging is the function of the centrifugal force experienced by its surface. The bulging with a C60 card, even at above speed, is again significantly lower. This enables very fine settings of the parts around the cylinder.

Loading on the cylinder, as previously described, is a very important criterion for deciding the carding intensity. The higher the loading, the lower is the carding intensity. This is because, with higher mass or loading on the cylinder, the total carding area per fibre is reduced. The wider width of 1½ m of the C60 card, as against 1 m width of a traditional card, provides higher carding surface, thus enabling the same mass to be spread over almost 50% more carding area. This means that, with the same card production, the thickness of the fibre layer is reduced by the same margin. In other words, with the same thickness of fibre layer as with low width card, it is possible to increase the card production on the C60 card by almost 50% (Figure 6.30).

As shown in the figure, for the same production level (say, 100 kg/h), the thickness of the fibre deposition on cylinder with the larger width of 11/2 m is very small (sky-blue colour), whereas with the smaller width of 1 m it is almost 50% more (sky-blue + violet colour).

In other words, for the same layer thickness, the wider width card can give 150 kg/h as compared to 100 kg/h for the smaller width card.

As regards the quality of yarn spun, the following bar chart gives a fair idea of various yarn properties spun from both C50 and C60 cards. The yarn imperfections (Figure 6.31) of C60 are almost comparable in spite of the fact that it gives 50% more production.

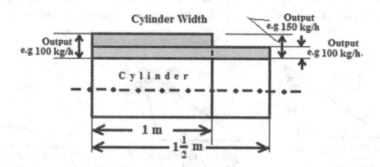

FIGURE 6.30   Fibre coverage on cylinder[5,6,8]: Increasing the width of the cylinder allows more area exposed for the fibres. This reduces the thickness of deposited fibre layer on the cylinder layer. Alternately, an increased cylinder width allows higher production rates by maintaining the same thickness of fibre layer on its surface.

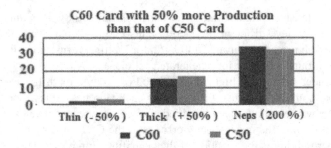

FIGURE 6.31 Yarn imperfections at higher production[5,6,]: On increasing card width, the reduction in the operational layer improves carding quality. The resulting enhanced sliver quality has marked influence on the yarn imperfections.

C60 is a universal card and can be used for all applications. So as to achieve the technical optimum, it is very important that the carding and extraction elements be replaced individually. In C60, it is possible to fit carding segments or cover segments at and near the extraction blade. With synthetic fibres, the use of carding segments serves a double purpose; they prevent fibre loss and achieve maximum opening action.

6.13.1.1   Rieter's Integrated Grinding System (IGS)[5]

As mentioned earlier, another important feature of the C60 card is IGS. During normal working, the card wires experience a constant wear and lose their sharpness. The rate at which the card wire wears depends on the class of work (whether relatively clean or more trashy cottons). Thus, when working on dirty cottons, the wires are expected to wear comparatively quickly. With conventional cards, the card is required to be stopped for carrying out grinding operation. Under the provision of IGS (Figure 6.32), the Rieters have equipped their C60 card with two types of automatic grinding systems: IGS-Classic (for cylinder) and IGS Tops (for flat tops). These two systems provide automatic grinding, which is carried out periodically without loss of any production. Thus, the sharpness of the wires is regularly and periodically maintained.

The relation between the sharpness of card wires (clothing), the neps and the trash removed is well known. As regards the neps, the sharper cylinder wires easily release the neps on to the flat surface.

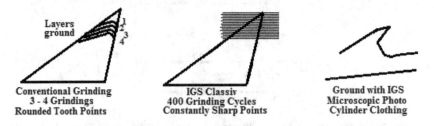

FIGURE 6.32   Conventional versus Rieter's IGS Classic Grinding[5,8]: In conventional grinding, the wire teeth lose their sharpness and become rounder after few grindings. In Rieter's IGS system, the sharpness of the tooth is maintained even after repeated grinding cycles.

The case with ejection of trash particles is similar. Whereas the fibres are held by the sharper wires, the trash particles are more effectively ejected. With IGS, the sharpness of the wire is preserved and hence the sliver quality is constantly maintained. It is claimed that, with IGS, the service life of both the cylinder and the flat wires is increased by about 30–35%.

The novelty of coupling card and draw frame in the C60 card is also unique. Normally, the card–draw frame coupling uses a creeping speed during can changing. However, a combination of C60 card coupled with draw frame offers can changing at a much higher speed. In short, the Rieter C60 card out-performs many other cards. It meets the highest quality standards for fine ring yarns and highest production performance for rotor yarns.

6.13.2 C70 CARD

Rieter's C70[5] card also incorporates a larger working area by increasing the width from 1 m to 1.5 m (Figure 6.33). Again at very high speeds of cylinder, the gaps around the working parts are very precisely maintained, thus ensuring the best possible quality. The specialty in this card is that an improved raw material utilization is achieved by providing adjustable knives around the licker-in. Inserts on Q-package and varying flat speeds further optimize the utilization of raw material.

Like the C60 card, a consistent sliver quality is ensured by the IGS system in the C70 card for the entire lifespan of the clothing. Owing to this, the service life of clothing is expected to increase up to even 20%. In C70 card, modular designs like C60 cards are available. These help in quickly replacing the modular structure (e.g., licker-in or its set) in quick time and there is considerable saving of the downtime of the machine. The faulty module which is taken out can then be conveniently repaired.

With an aero-feed system in the C70 card, it is possible to feed up to 10 cards with the total production requirements of 1200 kg/h. The aero-feed system ensures very uniform feed in the form of a sheet (bat) and this leads to a uniform card web.

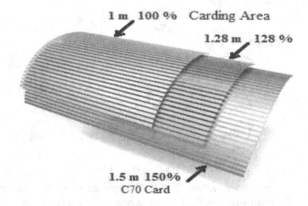

FIGURE 6.33 Active carding area[5,8]: Fibre to fibre separation or individualization of fibres is a very important requirement of the carding process. This largely depends on the effective carding area available to carry the job. Hence, the wider a card, the greater is the active carding area and hence improved carding action.

In C70, there are 32 working flats on the cylinder at any time. This gives a precise *Active Carding Index* (product of working flats and card width). Thus, whereas for 1 m width this index is 28, for the C70 card it is 48. With the accurate flat guidance system, it is possible to quickly provide for a precise carding gap (flat setting) depending on the material combination and the ideal tolerance.

With the C70 card, it is possible to optimize the trash removal. Depending on the cotton price and the extraction level of the trash, C70 offers considerable saving based on card line production. Both the manual as well as automated adjustments of mote knives around the licker-in are possible. The automated version is optional, but provides for changes in the setting during production. It is possible to set the extraction width (Figure 6.34) at licker-in, depending on the trash content in the material and the amount of waste required.

Another interesting feature of C70 card is the 'Q' Package provided in pre- and post-carding zones. In carding, it is always preferable to have a minimum proportion

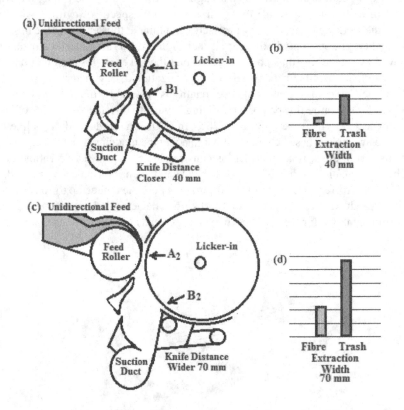

FIGURE 6.34 Waste extraction level[5,8]: The positioning of knife with respect to first striking point of licker-in wires on the lap fringe becomes very important. The larger this distance, more becomes the opportunity for the release of the trash; though there is some increase in the lint droppings. (a) Closely spaced knife, (b) more trash, but more lint as well. Distance—A_1 & B_1—extraction width. (c) Widely spaced knife, (d) lower lint, but lower trash. Distance—A_2 & B_2—extraction width.

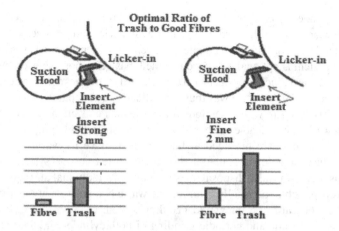

FIGURE 6.35   'Q' package in pre- and post-carding zones (C70 card)[5,8]: Q-package involves use of special inserts which help in achieving the optimal ratio of good fibres to waste. The optimal utilization occurs owing to use of low-wear mote knives in pre-and post-carding zones. The use of different designs of inserts help in this process.

of good fibres in the trash extracted. It must be also noted that whenever efforts are made to extract more trash, they are always accompanied by more fibre loss. In C70 (Figure 6.35), the concept of optimal ratio of trash to good fibres is the best and maintained at different extraction widths. It ideally balances both the fibre and the trash in the waste extracted at card.

Special pieces called *inserts* are provided for adjusting the distance (Figure 6.35). It is very easy to change the type of insert without using any tool.

The inserts are low-wear mote knives put in the pre- and post-carding regions. Different types of inserts—open, fine, medium and strong—are available for tackling varying degrees of contaminations.

Variable flat speed (Figure 6.36) is yet another feature of C70 cards. The use of a frequency converter enables continuous adjustment in the flat speed. This change

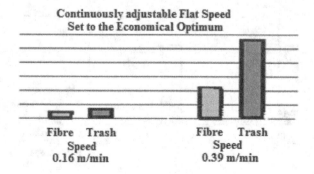

FIGURE 6.36   Adjustable flat[5,8]: The change in the flat speed is brought about by using a converter drive. It is known that with higher speed, the flats are expected to gather more trash on their wire surfaces. In this case, however, there is some increase in the lint, which the flats also hold.

can be independently carried-out and does not depend on cylinder speed. It is thus possible to exactly align the card to suit the raw material being processed.

With the C70 card, there is a unique arrangement of separate licker-in waste disposal. This system is optional but, once installed, has very short payback time. Much earlier, all the waste, including cleaner flat waste and dirtier licker-in waste, was combinedly removed. The mixture was thus less valuable. The separate removal and collection of dirty licker-in waste therefore, allows distinct collection of a comparatively cleaner flat waste. The saleable value of cleaner flat waste is thus appropriately realized.

The suction ducts are provided at licker-in to remove its waste intermittently. This waste disposal is further integrated with the waste transport system of the blow room. Thus, licker-in waste disposal does not have any additional air requirement. In addition, it is possible to visually examine the waste composition at any time.

Like the C60 card, the integrated grinding system (IGS) of the C70 card also incorporates automatic and periodic grinding of both cylinder and flat wires. Here, too, as many as 400 times the wire grinding is carried out during the whole service life of the wires to keep them absolutely sharp for the work. Based on this, the programme automatically calculates optimal distribution of *grinding cycles*. A comparative bar chart (Figure 6.37) gives the time required for the whole grinding operation.

The time for changing the licker-in, flats or even doffer clothing in C70 is greatly reduced. The spare modules are available and, therefore, it becomes quicker to replace the damaged module with a fresh one. The labour requirement for bringing about the replacement to its position is also bare minimum. Just one person from the maintenance staff is required to carry-out the job. For example, it is claimed that the licker-in module replacement can be done in just 1½ hours.

With C70, there is an improvement in performance of about 115% over the normal C50 card and an energy saving of up to 40%. Depending on the raw material, it is possible to achieve 1200 kg/h production with a line consisting of 10 cards. Rieter has unique system of *pressure control* in card chutes.

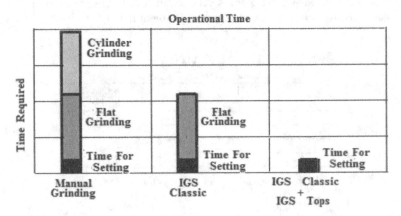

FIGURE 6.37 Time required for operations[5,8]: A manual grinding involves a lot of time in carrying-out flat grinding operations. Whereas IGS classic eliminates extra time for cylinder grinding, an added IGS tops equipment further get rid of additional time for flat grinding.

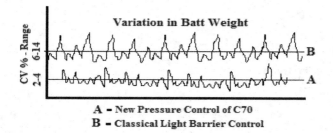

A – New Pressure Control of C70
B – Classical Light Barrier Control

FIGURE 6.38   Batt weight over 24 hours[5,8]: It is necessary to control the weight of the cotton material in chute feeding. M/s Rieter have a new device to control the pressure in their chute feeding system. The system meticulously controls the batt weight fed to the feed system of their C70 card.

With this system, it is possible to precisely control the batt weight during material feed to the card. Due to this, variations in the batt weight (Figure 6.38) are substantially reduced.

Cotton contaminated with honeydew often tries to develop sticky surfaces on material guiding elements. Similar is the case with synthetic fibres which contain strong lubricating fibre finishes. The deposition of such matters often creates problems in carding. Apart from web lapping in some instances, the held-back fibres are suddenly released and form a thicker portion in the condensed web. The only solution in the normal case is to clean such surfaces.

However, in many instances the cleaning becomes time consuming operation. A new web bridge (Figure 6.39) in Rieter's C70 card involves only the shortest possible time to remove, clean and reinstall the device. This bridge allows a perfect and uniform web and there is a sizable reduction in thick places. If there is an appropriate replacement of part, the down time can be still further reduced.

The actual sliver formation (condensation of web) consists of two aprons that reliably guide the delicate web and hence it is possible to produce a fine hank of the

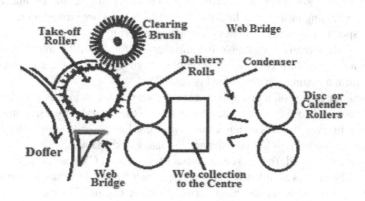

FIGURE 6.39   Web bridge on C70 card[5,8]: A new web bridge is a special attachment to M/s Rieter's C70 card. It ensures a perfectly uniform web. It is claimed that with this bridge, thick places are substantially reduced.

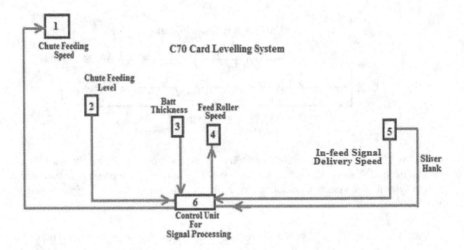

FIGURE 6.40 Levelling system[5,8]: The sensing devices at different stages in the card processing send their signals to the central processing unit. Even the signals are received from the outgoing hank of slivers. Finally the corrections are passed to chute feed and feed rollers to level the feeding at the licker-in side.

sliver. Especially with aprons guiding the web and subsequently condensing it, a sliver as fine as 4 ktex can be produced.

There is a provision in Rieter's C70 card to carry-out both short-term and long-term levelling. The former measures the thickness (actually the mass) of the fibre batt. This measurement is then automatically conveyed to the correction system which adjusts the speed of the feed roller. The corrected feed to licker-in thus ensures the uniform fineness of the card sliver.

As shown in Figure 6.40, there is a central processing unit which receives the signals from four different places: (a) material level in the chute feed (b) the material (batt) thickness as it is about to enter the feed roller system, (c) speed of the final delivery roller and (d) delivered sliver hank. These are judged and evaluated at the central processing unit. Accordingly the signal messages are passed on to: (e) chute feeding speed-unit which regulates the material entering the chute and (f) feed roller speed, which is really responsible for allowing uniform feed to enter the licker-in processing zone. Thus, once the instructions are received by (e) and (f), a constant and uniform amount of material feed is ensured.

In long-term control, a disc roller at the delivery side continuously measures the sliver fineness. The signals are once again sent to the central processing unit which compares them with the normal value and accordingly sends signals to the feed roller speed-control unit to make the appropriate speed changes.

Like the C60 card, the most economical feature of a C70 card is *process integration*. The card is coupled with the Rieter draw frame. This has been proven to be of great advantage for spinning yarns on a rotor system. The cleanliness of the sliver, along with parallelization of the fibres to a certain extent, are both essential requirements for rotor spinning and are automatically met in the C70 card system. The draw frame module, as mentioned before (a separate module), comes

FIGURE 6.41 Tongue and groove—C70 Card[5,8]: A powerful device that senses the variation in the sliver mass and makes the sliver more uniform by bringing about the necessary changes.

in two versions—with SB draw frame or with RSB draw frame, the former without auto-levelling and the latter with it. The attachment of the draw frame to a C70 card helps in reducing the draw frame passages in subsequent processes. The auto-levelling is done by typical tongue and scanning rollers (Figure 6.41). The scanning rollers are pneumatically loaded and this ensures complete scanning. The material thickness (mass) is measured and conveyed to digital processor which compares it with nominal value. Ultimately, the signals are sent to highly dynamic servo motor drive, when the precisely measured sliver portion reaches the levelling point. in the main draft zone. This guarantees complete levelling of the sliver material (Figure 6.42).

The C70 card has been proved to be beneficial. Together with a highly flexible blow room line, the sequence up to card is capable of adjusting to process blends of cotton and synthetic fibres simultaneously.

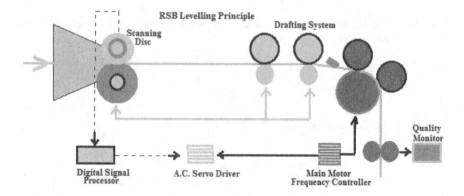

FIGURE 6.42 Levelling principle[5,8]: In RSB auto-levelling principle, digital signals from the scanning disc are processed and passed on to the servo driver. Accordingly, the draft in the draw box is adjusted to maintain the mass per unit length.

This is all made possible because of a modular design wherein fibre guidance components are specially made of chromium. The clothings of the licker-in, cylinder, flats and doffer are specially designed to suit processing of synthetic fibres. The carding segments are also specially made to suit the purpose.

The typical technical features and production data for C70 cards include the following:

- The card is able to process both cotton and synthetic fibres (with quick changes).
- It can process a very heavy batt weight (up to 900 g/m) and is capable of giving production up to 280 kg/h.
- Sliver fineness varying from 4–20 ktex can be produced.
- With cylinder speed varying from 600 rpm to 900 rpm, a sliver delivery speed of 30 m/min is possible.
- With a central suction device for separating licker-in waste and flat strip, it is necessary to have an exhaust air speed of 1.2 m³/s, and the compressed air requirement of 0.7 Nm³/s. With this, the total installed power requirement is 21.3–29.6 kW.

REFERENCES

1. Manual of Cotton Spinning – "carding" – W.G. Byrerly, J.T. Buckly, W. Miller, G.H. Jolly, G. Bettersby, Frank Charnley, The Textile Institute Manchester, Butterworths, 1965
2. Spun Yarn Technology – Eric Oxtoby, U.K. Butterworth publication 1987
3. Elements of Cotton Spinning, Carding & Draw Frame – Dr. A.R. Khare, Sai Publication, 1999, Mumbai - 400078
4. Process Control in Spinning - ATIRA Silver Jubilee monographs, 1974
5. The Rieter Manual of Spinning, Vol. 2, & Book-lets
6. Trumac/Trutzschler pamphlets, brochures, book-lets etc.
7. Technology of short staple spinning – W. Klein, Textile Institute Manual of Textile Technology, 1987
8. A Practical Guide to Opening & Carding - W. Klein, Textile Institute Manual of Textile Technology, 1987

7 Faults and Their Elimination

7.1 NEP COUNT OF THE WEB UNSATISFACTORY[1,3]

The neps are formed in carding when the fibres go out of control and are allowed to roll during processing.

7.1.1 CONDITION OF WIRES

If the last grinding applied to the wires was done long time ago, the wires become dull and lose sharpness. A fresh grinding, therefore, is necessary. If the wire itself has been in use for a long time, it really becomes old. In this case however, it needs to be replaced.

7.1.2 LOADING ON THE WIRE SURFACES

When the cylinder wire surface becomes overloaded, the carding action suffers and as a result, neps are produced. It is possible that the wire surface where the loading is predominant is damaged or there is accumulation of some sticky dirt.

7.1.3 SETTINGS

The back plate setting usually controls the distribution of fibres on the cylinder. If air currents around the cylinder in this region are not guided properly, they lead to turbulence and this allows uncontrolled movement of fibres, thus leading to formation of neps. The action of the flat is very important in removing neps. Both their speed and the setting with cylinder decide the nep removal potential. A closer setting between cylinder and doffer helps in effective transfer of fibres. A poor transfer in this region implies that the fibres are allowed to go around the cylinder repeatedly and this increases the nep level in the card web.

7.1.4 IMMATURE FIBRES

If the raw material contains a high proportion of immature fibres, it leads to high nep generation. This is because these fibres can curl, bend or roll quickly to form clusters. Especially with mechanical harvesting, it is highly likely that some immature cotton balls are picked-up along with the normal lot. In ginning, these fibres give rise to nep formation.

DOI: 10.1201/9780429486562-7

7.1.5 Production Rate

A higher production rate through higher doffer speed or lap weight always leads to higher nep generation (except in a modern high-production card).

7.2 HOLES IN THE WEB[1,3]

A hole in the web means a very thin portion. When it is immediately followed by a lump of material, it means that at some point on the cylinder undercasing, the material has been temporarily held and then released. Checking the joints of two undercasings or any roughness on its working surface reveals the source.

If either cylinder or doffer has a damaged portion of blunt wires, the result will be occasional blanks in the web. Improper stripping of doffer by the comb owing to roughness developed on the comb blades also leads to formation of holes in the web. A light polishing by emery can be a remedy. A damaged blunt wire on the doffer is not in a position to receive fibres from the cylinder and also causes a blank or hole in the web.

7.3 SNOWBALL FORMATION[1]

When small balls come out with the web, the defect is termed *snowball formation*. When the fibres accumulate at the cylinder undercasing they are picked up by cylinder wires in bunches and are usually allowed to pass as balls. The accumulation is basically due to rough surfaces at the undercasing and these must be smoothed out. The flat portion of the cylinder undercasing (middle portion) should sharply bend downwards so that fibre accumulation is prevented.

7.4 CLOUDY WEB[1]

A web consisting of unopened fibres in general or clusters of fibres emerging together is direct evidence of poor carding. The web appears cloudy mainly because of unopened material appearing in patches against the semi-transparent web. This is due to blunt or damaged wires of the flat and cylinder, thus hampering effective fibre-to-fibre separation. The cloudy web is also due to a wide setting between cylinder and flats, and improper back plate setting (setting too wide or too close), which leads to air turbulence.

7.5 RAGGED SELVEDGES (WITH LAP FEED)[1]

Uneven and weary selvedges are the result of improper setting of selvedge lap guides. These guides should be adjusted to give a quarterfold to the lap before it is gripped by the feed roller. If the lap width is smaller than the card width, wooden packing of suitable width should be provided at the lap stand to keep the lap correctly at the centre. Damaged wires at the edge of the licker-in, cylinder or doffer may also result in ragged selvedges.

7.6   FLUFF GENERATION FROM THE CYLINDER SIDE[1,2]

When considerable fluff is generated between the cylinder side and the side framing that support the flexible bends, it is owing to a wide gap between them. Especially in high production carding, when the cylinder speeds are higher, this wide gap allows an escape route for air currents, which also carry along with them, this fluff through the sides. Therefore, it is essential that these gaps be suitably sealed. Improper squaring of cylinder with respect to the framing also leads to this type of defect. A gauge between 0.635 mm (25/1000 in) and 0.864 mm (34/1000 in) is found to be satisfactory to avoid excessive air leakage.

7.7   SHIFTING IN WEB STRIPPING POINT[1]

In conventional cards using doffer comb, the web should detach itself from a fixed point just below the sweep of the doffer comb. However, when the doffer wires become dull and blunt because of improperly scheduled grinding, the point at which the web leaves the doffer shifts up and down around the doffer comb. Though it cannot be truly termed as a defect, it certainly indicates the need to improve the doffer wire condition.

7.8   WEB SAGGING[1,3]

In the conventional card, the web, as it leaves the doffer, slightly curves around the doffer comb and then moves further along a straight path towards the nip of the calender roller. However, when the major portion of the web sags low, it is due to different reasons.

When the cotton is very soft and humidity is very low, the web starts sagging due to the poor clinging power of fibres. The lower humidity also induces static generation, which further aggravates the situation and sometimes even allows dropping of the part of the web portion.

When the tension draft between calender roller and doffer is very low, the web may sag because of insufficient pull from the calender roller. Very high humidity, on the other hand, makes the fibres in the web absorb excessive moisture and thus the fibres are in a wet condition and become heavier. This may also lead to sagging of the web. Raising the mean oscillating position of the doffer comb was one of the remedies sought in the mills with conventional cards.

7.9   POOR CARD CLEANING EFFICIENCY[1,3]

After the blow room, carding is perhaps the only major process for cleaning the remaining trash from the cotton. Therefore, cleaning efficiency at the card has an important bearing on the quality of the sliver produced.

In card, the major cleaning takes place in the licker-in region. Hence, for very effective removal of trash, the licker-in wire condition must be really good. The licker-in wire points for this purpose must be sharp enough to act very positively on the lap fringe presented to it. The speed of the licker-in, contrivances around it and the type of wire all play an important role in this action. The feed rollers must move smoothly and uniformly. The pressure must be adequate so as not to allow any snatching and at the same time should ensure uniform feeding of material to the licker-in.

The proper condition of the mote knives (their sharpness) and the correct settings help in giving the best possible cleaning. The correct setting of licker-in undercasing, as well as the proper choice of the type of undercasing, can also increase the trash extraction in this region. Attachments like comb bars or deflector plates have proved their benefits in improving both the opening as well as cleaning at licker-in in conventional card. In a modern comber, there are deflector plates, sharp knives and suction hoods to immediately extract the impurities released.

The right selection of type of flat tops and their action on the fibres also helps in extracting certain types of foreign matter such as seed coats, leafy vegetable impurities etc. In this connection, both the flat speed and their setting with the cylinder are important. A complete cleaning of the flats when they move away from the cylinder should not be overlooked. This is because it is equally important to present them fully fresh and cleaned when they return their journey over the cylinder from the other side for continued carding action.

The dust extraction units at the card play an important role, too. The fly-fluff and fine dust are sucked by this unit and hence, the screen over which these are deposited must be cleaned at regular intervals (with old conventional cards). The connection of the hose pipes must be properly sealed so as to avoid any air leakage.

With high production carding, the role played by cross rolls in pulverizing the thorny impurities is important in allowing the same to fall down from the web. It is, therefore, essential to maintain the required pressure on these rollers.

A correct level of humidity in the card room is yet another factor which requires attention. With higher humidity, the fibres in the lap acquire more moisture and the extraction of impurities at the licker-in region is seriously affected. This is because with more moisture, the foreign matter adheres more tenaciously to the fibres and their separation at the licker-in zone or elsewhere becomes difficult. Usually the trash extracted in such cases is usually accompanied by higher lint loss.

The higher trash content in the lap also leads to higher trash in the sliver (Table 7.1). Though card cleaning efficiency is higher with trashy laps, the content left over in the sliver is comparatively higher. Like the blow room, the card also adjusts its level

TABLE 7.1

Trash in Lap/Sliver and Card Cleaning Efficiency[3]

Trash % in Lap	Trash in Sliver	Cleaning Efficiency
0.7–1.0	0.2–0.3	70%
1.0–1.5	0.3–0.4	72%
1.5–2.0	0.4–0.5	74%
2.0–2.5	0.5–0.55	76%
2.5–3.0	0.55–0.6	79%

of performance according to the trash content in the lap when all other parameters are constant (Table 7.1).

In a particular study on a card, the results obtained clearly showed that, with higher trash content in the lap, the card cleaning efficiency is higher. It must, however, be borne in mind that an efficiency of around 79–80% seems to be poor, even when the lap is very trashy. It perhaps speaks for the condition of the machine. With standard condition and best maintenance, the expected cleaning efficiency is around the values shown in Table 7.2.

A combined action of conventional blow room and card tries to attain a cleaning efficiency of around 85–88%. In the modern blow room and card sequence, the efficiency level has almost reached 96–98%. With conventional card, it is observed that with a higher cleaning efficiency at blow room, the cleaning efficiency at card is somewhat lower and vice-versa. It is always an intriguing question as to whether to load the blow room more for attaining higher cleaning efficiency or allow the card to reach its best. This is because if the card is loaded more (more trashy laps), then the life of the card wires is considerably reduced. However, if the blow room action is intensified to achieve more cleaning, it is always accompanied with higher lint loss. The solution should, therefore, be economically viable. Usually it is advisable to allow the blow room to perform a little better in extracting trash, without allowing the lint percentage to increase significantly, and then leave the card to attain its natural performance level. With modern blow room, it is quite possible. The modern card is, however, capable of giving a far better opening as compared to blow room and hence is in a better position to reach a higher level of cleaning.

It is also possible to make efforts in modern carding to save the lint by appropriate adjustments around the licker-in region. It is equally possible to intensify the licker-in action to improve cleaning efficiency at card. The carding segments in modern card not only contribute to improved carding (individualization) but also improved cleaning. Further, it may be emphasized that the timely replacement of licker-in wires not only maintains the satisfactory cleaning level in these zones, but also reduces the load on cylinder wires considerably. From an economic point of view, too, this sounds to be a good proposition, as replacing licker-in wires is far less costly than changing cylinder wires. A good spinner thus should maintain the quality of various carding parameters at their optimum level so as to get the best functioning of a card as a whole.

TABLE 7.2

Trash Content and Expected Card Cleaning Efficiency (Conventional Card)[3]

Trash in Lap Fed	Expected Cleaning Efficiency
2.5% & more	78–80%
1.5%–2.0 %	76–78%
1.0%–1.5 %	73–76%
Less than 1%	70–73%

REFERENCES

1. Elements of Cotton Spinning, Carding & Draw Frame – Dr. A.R. Khare, Sai Publication, 1999, Mumbai
2. Spun Yarn Technology – Eric Oxtoby, Elsevier Publication, 1987, Butterworth.
3. Process Control in Spinning – ATIRA Silver Jubilee Monographs, 1974
4. A Practical Guide to Opening & Carding - W. Klein, Textile Institute Manual of Textile Technology, 1987, Manchester

8 Draw Frame

8.1 OBJECTS

Though in carding the tangled mass of cotton fibres is opened out to make them quite free—almost in fibre-to-fibre state—their arrangement in the card web is by no means uniform (Figure 8.1a). The fibres lie in a haphazard manner and are criss-cross. Such an arrangement can be termed as *random*. However, it is important to remember that it is this disposition of the fibres that enables the web to hang on to itself, almost unsupported, between doffer and calendar roller in carding. Had the fibres been parallel to one another and very uniformly arranged, their holding power on their neighbours would have been too weak to allow them to be doffed satisfactorily from the doffer in the form of web and then transported over a comparatively wider distance (in conventional card) to the calendar roller. Besides in this random arrangement, the fibres are hardly straight and their ends (either one or both) are bent to form the hooks. It can be easily appreciated that when the fibres are all parallel to one another and also to the sliver axis (Figure 8.1b) in a strand, each contributes its strength in the same direction. In short, their efforts are united.

The criss-cross arrangement is shown in Figure 8.1a. The inclination of each fibre varies anything from 0° to 90°. The fibre extent of each fibre along the sliver axis thus changes (Figure 8.1c). So, the contribution of each fibre along the sliver axis becomes different. For example, if the fibre is inclined to the sliver axis at angle Ø and if C is its contribution along its length, then the component of this contribution along the sliver axis AB (Figure 8.1c) will be only [C. cos Ø]. Therefore, the fibres must be straightened and parallelized to both, themselves and to the sliver axis.

When the hooks are straightened-out (Figure 8.1d), a much greater length of the fibre is available along the sliver axis and hence along the length of final yarn. Both the fibre parallelization and its straightening improve the fibre orientation; otherwise the fibres would be treated as much shorter than their actual length. It may be mentioned here that a classical relation exists between the effective length of the fibre and the strength of the yarn spun from it. Therefore, the greater the length of the fibre realized along the material axis (Figure 8.2), the higher is the strength of the strand. The fibre parallelization and orientation are thus important in this respect.

Another important object of the draw frame is to improve uniformity of the strand. Though the card sliver may be fairly uniform over a short length, it is not so when longer lengths are compared, i.e., it also has medium- and long-term length variations. Therefore, in draw frame, the efforts are directed to make the sliver uniform and even. In short, the objectives of the draw frame can be summarized as follows:

1. To straighten out the fibres and to improve the fibre extent.
2. To parallelize them to their neighbours and to the sliver axis so that in the final stage of spinning, they can make the maximum contribution towards the yarn strength.

DOI: 10.1201/9780429486562-8

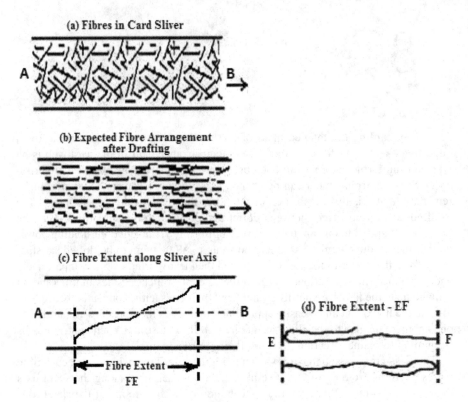

FIGURE 8.1 Fibre arrangement in card[1,2]: It is known that the fibre arrangement in the card sliver is most random. The fibres are criss-cross and hooked. Especially when the fibres are hooked, their extent is shortened. The process of drafting improves this and brings the fibres more aligned to the sliver axis. (a) Fibres in card sliver, (b) expected fibre arrangement after drafting, (c) fibre extent along sliver axis and (d) fibre extent.

3. To improve the uniformity and evenness so that the sliver thus produced becomes more regular.

In draw frame, these objectives are met with two operations—drafting and doubling. The drafting operation means drawing or pulling-out (Figure 8.3) the strands of fibres. A group of card slivers of given bulk are drawn into one single

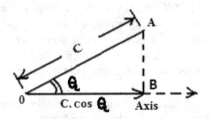

FIGURE 8.2 Contribution of fibre properties[1,3]: When the fibres are not parallel but inclined to the material axis, the contribution of their useful properties to the material formed gets affected.

Thinning-out Operation

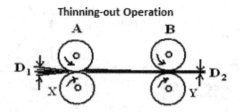

FIGURE 8.3 Thinning-out operation[1,2]: When the material passes through the pairs of top and bottom rollers running progressively faster, the material is stretched and becomes thinner.

strand of much greater length but of equivalent bulk of the individual original sliver. As shown, A and B are two pairs of rollers. If their surface speeds are the same, the material entering at X and will go through the system without any change to come out at Y. However, if the surface speed of pair B is twice that of pair A, the material (D_1) entering the system will be reduced in such a way that $D_2 = \frac{1}{2} D_1$. However, the weight of the material entering the system per unit of time remains the same as that leaving it. Thus, if ℓ is the length fed during a certain time, then the length coming out of the system, during the same time, will be 2ℓ in this case.

At pair A, the material (D_1) entering the system will be reduced in such a way that $D_2 = \frac{1}{2} D_1$. However, the weight of the material entering the system per unit time remains the same as that leaving it. Thus, as mentioned above, if ℓ is the length fed during a certain time, then the length coming out of the system, during the same time, will be 2ℓ.

In fact, this is called the *drafting operation*. In blow room and carding, there is drafting or thinning out of the material, but only as an accessory to the main objectives of attaining opening, cleaning, disentanglement, etc. In blow room, the cotton is pulled out (drawn) from the bulk, but the objective is never to draw the material out. In carding, there is an individualization when the material is thinned out; again, the objective is never to draw out the material. On the other hand, subsequent to the draw frame operation, in speed frame and final ring frame the drafting is done more to thin out the material so as to make it finer in weight per unit length. Only in draw frame is the objective mainly to draw out the material into a longer length in order to attain parallelization of fibres. Here, the weight of the material per unit length—fed and that delivered—is more or less kept constant by yet another process called *doubling*.

It is interesting to note that the drafting on draw frame implies straightening and parallelization of the fibres. Therefore, if the slivers entering the system (Figure 8.4) have the fibres criss-cross and randomly disposed, the pulling action of the roller pair B automatically straightens and aligns the fibres in the slivers along the direction of pull. Simultaneously, as the fibres under the influence of pair B are pulled at a faster rate (surface speed of B being assumed to be twice that of A), their rear ends in finding their way through the tangled mass are predominantly straightened-out.

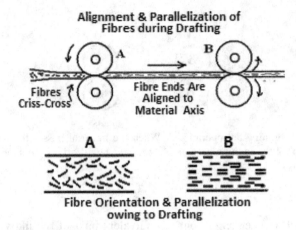

FIGURE 8.4   Fibre alignment[1,2]: The process of drafting improves both fibre orientation and fibre alignment to the material axis.

Thus, while the fibres are being parallelized to the sliver axis, the hooked fibre ends are also straightened out. It has been accepted that during pulling of the fibres, their rear ends are more favourably straightened. Even then, the hooks in the leading direction are also partially removed. This appreciably improves the fibre extent further (Figure 8.5).

Doubling implies feeding more than one strand of sliver to the system, e.g., as mentioned in the earlier case, if two slivers are fed to the system of two pairs of drafting rollers A and B (Figure 8.6) with draft of 2 (ratio of surface speeds B:A: : 2:1), each one will be reduced to half of its weight, whereas the slivers will be stretched or lengthened to twice their original length. Ultimately, under this condition, the

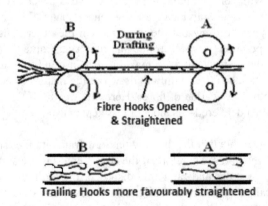

FIGURE 8.5   Straightening and removal of hooks[1,2]: When the fibre strand is drafted, the fibres in it are straightened out. This also leads to partial removal of hooks.

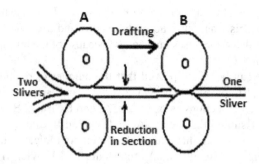

FIGURE 8.6 Maintaining weight of outgoing sliver[1,2]: With a certain value of draft and the chosen number of doublings, it is possible to make the sliver more even and also conform to the same individual sliver weight per unit length fed.

emerging sliver has almost* the same weight per unit length as that of the average weight of the two combined slivers fed. Thus, with length fed as ℓ for each of the slivers, the corresponding length delivered through B will be 2ℓ. Owing to doubling, there is an improvement in sliver evenness and regularity.

8.1.1 EVENNESS

Evenness means uniformity in cross-section along the length of the material. The faults associated with unevenness of the final yarn are plainly noticeable to the naked eye. When measured at a very short length (less than 2 in), the thickness of the uneven yarn can vary as much as ± 50%. This pattern is often repeated in such a fashion that every thick place is followed by a thin one.

The important characteristics, such as strength and elasticity (extension% at break) of the yarn are invariably recorded at the thinnest and weakest places. Thus, evenness affects not only the appearance but the whole value and usefulness of the material—the final yarn.

8.1.2 REGULARITY

With reference to sliver, *regularity* means the uniformity in its hank. Whereas evenness is concerned with more immediate short-term variation, regularity refers to long-range variations. Therefore, it is a true measure of sliver or material coverage at any stage.

The regularity depends on combining (doubling) more slivers together (in old conventional machines this was 6; in modern draw frames it's 8 or 10). It is precisely because of the positive contribution of doublings in levelling up the irregularities that comparatively fewer doublings are enough to bring a much higher degree of

---

* Depending on the draft employed and final hank of the sliver decided.

uniformity in the weight per unit length of the sliver. If the doubling operation is continued beyond this stage, it is not only useless and a waste from an economic point of view, but also it increases unevenness because of the multiplicity of drafting waves (see section 8.7.1).

Statistically, it can be easily proved that the number of doublings can improve regularity. If we just consider only two slivers, A and B, each six meters, we can think of six imaginary sections as $A_1, A_2,..., A_6$ and $B_1, B_2,..., B_6$, etc.

Let it be also assumed that except A_1 and A_4, and B_2 and B_5, all other sections in the two respective slivers have normal weight. As indicated in Figure 8.7, sections A_4 and B_2 are 12% lighter and A_1 and B_5 are 12% heavier. If these two slivers are combined (or doubled) into one thicker sliver, the thickness of the resultant sliver will depend on the relative positioning of each of the imaginary sections of the respective slivers. As per permutations and combinations, there are $6 \times 6 = 36$ possible ways in which A can be placed opposite each of the six sections of sliver B. Thus,

      (a) Chances of 0% variations   18 ways
      (b) Chances of ±12% variations   2 ways
                      Total   36 ways

The slivers which had ±12% variation in the original sliver pieces covered 1/3 of their original length. This means that each of the original slivers had 33.3% of their portions either heavier or lighter by as much as ±12%. However, after the permutations and combinations, the resultant sliver thus formed will have only 5.55% chances of having the same variation (2 ways out of 36 possible combinations). As against this, 66.66% of the original portion of each of the sliver had 0% variation (4 sections in every six sections), whereas the same is now reduced to 50% (18 ways in total of 36 possibilities).

One can see that though there is a possibility that the percentage of the original normal section of each of the slivers is reduced when the two slivers are combined, the possibility of the portions having extreme variation of ±12% variation is also

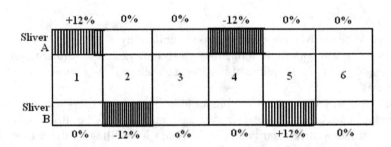

FIGURE 8.7 Improvement in regularity through doublings[1,2]: The process of doubling makes the sliver, roving and yarn more regular along their length. Even statistically it can be proved.

substantially reduced. In short, it can be observed that the magnitude of extreme variations is reduced and spread over a greater length of the material, thus reducing its impact. Instead of two slivers, if six similar slivers are doubled, the possible combinations lead to chances of variations as follows:

Chances of 0% variation	13,236 times
Chances of ±12% variation	2 times
Total	46,656 times

[**Note:** $6^6 = 46,656$]

In a conventional draw frame, where six slivers are customarily doubled, the possibility of actual reduction in variation would be somewhat like shown above. It can be seen from these two examples that the chances of original high variation of ±12% occurring in the resultant product are really remote. In fact, a closer look at these two examples would reveal a very important criterion of the influence of the doublings in suppressing the variation. As can be imagined, if the amplitude of the variation (curve P) is NX in the feed slivers (before doubling), it is expected to reduce to MX_1 (curve Q) after doubling (Figure 8.8). The variation in the sliver delivered by the draw frame is thus significantly reduced.

However, it does not mean that the variation disappears totally. This is clearly indicated in the graph. The process of doubling, therefore, is able to only reduce the amplitude of the wave originally present in the feed slivers. It means that when the slivers are doubled, the impact of severity due to variation is only reduced.

However, reducing the impact of the defect in this manner has its own repercussions. This is because the fault, once present in the material, cannot be removed totally.

In carrying out the doublings on draw frame, this is achieved by spreading out the irregularity over a greater length. As the sense of irregularity to the naked eye is restricted only when a smaller length of material is viewed at a time, the spreading of the irregularity over a longer length and reducing its magnitude lessens its significance to human observation.

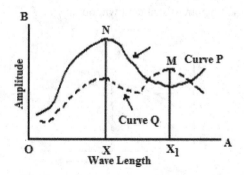

FIGURE 8.8 Effect of doubling on reduction in amplitude of variation[2,3]: The process of doubling mainly reduces the peaks or extreme variations in the material.

With the doubling process, therefore, there is a chance of thick and thin places present in the feed material (slivers) getting well blended together with those portions originally correct in weight. The chances of the level of original variations present in the delivered sliver are thus negligible, particularly when two such draw frame passages are employed. In fact, after the second passage of the draw frame, the statistical probability of occurrence of the original extreme variation will be—only 1 in—a number having 28 digits. Therefore, it can be safely ignored. In actual practice, however, it does not appear to be so simple (See section 8.7).

The amount of draft employed in a particular situation depends on the design of the drafting zone, the fibres' length distribution and the fibre extent.

As a general rule, more draft may be employed to those cottons with low CV% of fibre length. There is always an optimum draft (or range) for obtaining the best performance in terms of sliver uniformity. The level below and higher than this optimum (Figure 8.9) will always lead to lower uniformity of the product. It is observed that this optimum range depends on the type of fibre processed and also the machine used (the drafting system). When the higher draft must be used, there needs to be a higher accuracy in the design of the machine, especially the drafting system.

8.1.3 SHORT AND LONG FIBRES

It is well known that any type of cotton has a fair distribution of different lengths of fibres. There are some fibres which are very long and a few fibres which are very short. But a majority of them have a length around a specific value—the *effective length*.

Ideally, in a strand of material (sliver), all the above categories should be fairly represented. This point is extremely important, although not often considered so, and sometimes it is overlooked altogether. If the fibre distribution is faulty, it is seen in the resultant strand or the final yarn that at some cross-sections there is greater proportion of shorter fibres, and at some other places longer fibres dominate. Such a strand or a yarn is always weaker at those places where longer fibres are insufficiently represented.

However, no machine is able to offer a perfect distribution. The best that is possible is distribution of the large mass of the material in a random manner so that the various lengths of the fibre are distributed without any bias. This is best done at the

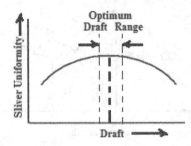

FIGURE 8.9   Draft anduniformity[2,3]: There is an optimum value for draft where the sliver uniformity is the best. Employing draft below or higher than this is not beneficial.

card. However, draw frame does it satisfactorily. In spite of this, the random distribution still deviates from the ideal in finished yarn. Therefore, it is necessary to control the uneven representation of short and long fibres across any cross-section and also along the length of sliver; otherwise, this fault will cause a lot of trouble in obtaining a uniform product: yarn.

8.1.4 RELATIVE POSITION OF FIBRES

The relative placement of fibres is another important condition. Even with the perfect distribution of short and long fibres and the constant weight per unit length the sliver and hence the ultimate yarn formed, can, by no means, be satisfactory. Their relative the sliver, and hence position in the strand (Figure 8.10) is very important. Especially in the yarn, the fibres need to be placed so that as few of them as possible begin or finish their ends in any one section of the strand.

Assuming that there are 100 fibres in a strand and each has a 1 in length, then ideally each fibre should be placed 1/100 of an inch ahead of the preceding one so as to have their perfect placement. Such a strand is bound to have at least 99 fibres at any place when the strand is cut across its width. The strength of the yarn is directly influenced by this condition.

It is because, at any point, the fibres, which are cut through an imaginary section, actually would contribute to the yarn strength at that point.

Consequently, the maximum strength is attained when the fibres are staggered longitudinally at perfect and regular distances. However, here again, it is very difficult to obtain such a distribution; the results, if obtained, would be ideal, though.

One of the reasons why shorter fibres make a weaker yarn is that there is a far greater number of fibres ending at any point. As shown in Figure 8.10, only C gives such ideal positioning of the fibres.

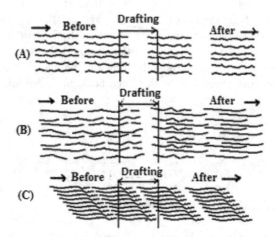

FIGURE 8.10 Relative position of fibres in a strand[2,3]: The concept is only hypothetical, as no machine can arrange the positions of the fibres so flawlessly. However, the idea is more related to equal number of fibres in the cross-section at any point in the ideal material.

In the remaining situations, the effect of draft only aggravates the gaps and disturbs the fibre distribution that already is not ideal in the feed material. The distribution 'C' is ideal because it maintains the uniform number of fibres in the cross-section of the material, before and after drafting.

8.2 DRAFTING BY ROLLERS

On a conventional draw frame, there are four pairs of drafting rollers (Figure 8.11). After the slivers enter the nip of the back pair A, they pass through the corresponding nips of the succeeding pairs, B, C and D, to come out at K.

Each pair of rollers is made to run progressively faster than the preceding one. The maximum difference in their surface speed is between pairs C and D, whereas the minimum is between pairs A and B. The ratio of the surface speeds between each pair is termed *draft*. For example,

$$\text{Draft between B \& A} = \frac{\text{Surface speed of B}}{\text{Surface speed of A}}$$

Where,

Surface speed of A = π × Diameter of A × rpm of A
Surface speed of B = π × Diameter of B × rpm of B

The ratio of surface speeds of D and A gives the *total draft* in the system. The ratio between pairs D and C is called the *main draft*, whereas that between B and A is called the *break draft*. In a conventional drafting system, the draft between pair C and pair B lies between the values of the main draft and break draft.

Thus, the values of the drafts from back to front (from A to D) progressively increase and the drafting system of this kind is again called a *graduated drafting system*. Pair A, where the material (slivers) enters the system is the *back roller* (BR). The following rollers are the *third roller, second roller* and *front roller* (FR), respectively. Thus, a system of 4 over 4 drafting automatically leads to three zones: the back or break zone (between the third and back pairs), middle zone (between

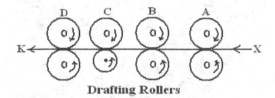

Drafting Rollers

FIGURE 8.11   Drafting on conventional draw frame[1,3]: Normally there are 4 pairs of drafting rollers and they run progressively faster. Depending on the number of slivers entering the back pair nip, the amount of draft is adjusted. Accordingly, the rollers are also set gradually closer. Such a system is called as *graduated drafting* system.

the second and third pairs) and the front or main zone (between the front and second pairs). As mentioned earlier, there are six slivers fed to the drafting system where they are progressively thinned out or attenuated (or drafted) in such a way that the final sliver is not very much different in its size, i.e., in weight per unit length, from the individual slivers fed. Therefore, the draft employed in the system is also six (or around six).

It may now be realized that corresponding top and bottom rollers (Figure 8.12) form a pair in a drafting system. The work of each pair of rollers can be considered satisfactory if the fibres passing through them travel precisely at their surface speeds. If this were not the case, the amount of draft in each zone would vary, and consequently, the hank of the material (sliver) would fluctuate. Hence the top rollers, in addition to their own weight, are further weighted to ensure their correct contact with their bottom counterparts to give a precise delivery speed. Incidentally, all the bottom rollers are positively driven through the machine gearing, whereas the top rollers merely resting on the bottom rollers (with additional weighting) are made to run by the surface contact with the latter. In this context, it can be easily understood why the top rollers require additional weighting on and above their own weight.

In fact, in proportion to the additional weighting (or pressure) that is employed, the gripping between the top and their corresponding bottom roller improves. The additional weighting (or loading) put on the top rollers necessitates strong bottom rollers which can take this pressure. They are, therefore, made of steel. Obviously, in order to bear a certain weight (or pressure), they need to have a certain minimum diameter (along with improved metallurgy). This is because, a value lower than this, makes the bottom rollers too weak to withstand the top roller pressure. Eventually, in such a case, the rollers are likely to buckle. In addition, the bottom rollers are fluted so as to increase the overall gripping action. The top rollers are not fluted but are covered with special synthetic rubber. Thus, when each pair of drafting rollers having synthetic rubber-cushioned top rollers and fluted bottom rollers is additionally loaded with a mechanical or pneumatic device, it gives much improved gripping action.

Though the movement of the fibres through a drafting system must be controlled, they must freely slide when the faster-moving pair tries to pull them. It means that when the fibres come under the influence of the next pair of rollers, their trailing

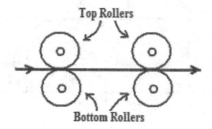

FIGURE 8.12 Top and bottom drafting roller pairs[1,4]: The characteristics of both the types of rollers are different. Whereas the top rollers are covered with soft cushion, the bottom ones are hard, metallic and fluted.

ends must be released from the grip of the preceding pair. For this, the rollers are set at a certain distance. The condition here must be such that at no time, both the ends of the fibres are allowed to be gripped by succeeding drafting roller pairs. Otherwise the drafting action between the two pairs would result in the back pair trying to hold the trailing end of the fibre at the same time the front pair trying to pull the leading end of the same fibre. This may put a lot of strain on these fibres and eventually break them or at least notably weaken them. Alternately, it may also lead to roller slip. Thus, it puts a great limitation on the use of this type of drafting system. As can be seen from a Baer sorter diagram (Figure 8.13), in general, the number of fibres in different length groups varies greatly. There is a certain percentage of fibres having length longer than the effective length. Though much less in proportion, they must be considered when carrying-out the roller setting.

Thus, the setting (deciding the nip distance) between the pairs of any two succeeding drafting rollers must be wide enough so as to protect these longer fibres.

In actual practice it is the *effective length* (EL) which is commonly referred to and considered when setting the rollers. It is observed that these longer length fibres are greater than effective length by a certain margin. Therefore, a calculated allowance is given on and above the EL to arrive at a value of setting between any two consecutive pairs.

When the fibre lot contains a substantial proportion of fibres shorter than the effective length, the wider setting, which otherwise is necessary for protecting longer fibres, puts these shorter fibres much out of control (section 8.7.1) and they become unmanageable during drafting. Only when all the fibres are of the same length, it is possible to arrive at a satisfactory setting value.

The ability to control all the fibres and yet allow free movement to some of them when required decides the efficacy or capability of any drafting system. Incidentally, the conventional 4/4 graduated drafting system has a poor capability to do this. Besides, when the difference in the surface speeds between the adjoining two pairs of rollers is increased (i.e., higher draft), the inability to exercise control over the fibre movement becomes more pronounced. There is also a considerable effect on the

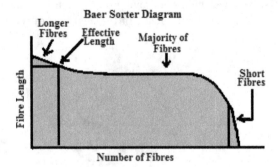

FIGURE 8.13 Baer sorter diagram[1,2]: The diagram shows the fibre distribution in a cotton lot and gives a fair idea of various properties, such as proportion of longer and shorter fibres. One of the properties (effective length), as explained above, is useful for arriving at the ideal roller settings.

quality of the product delivered owing to the speed of the production, a higher speed further deteriorating the quality.

In short, it becomes essential to exercise more control over the fibre movement when increasing the drafting capacity or production rate. It may also be noted that in graduated drafting systems, as there is a progressive increase in the surface speeds of pairs of drafting rollers from back to front, there is a progressive reduction in the mass held under each pair. It means that the mass worked over in each zone, from back to front, also reduces.

During drafting, when the fibres are pulled through a bunch of neighbouring fibres, they experience resistance. Therefore, a higher force (drafting force) is required to overcome this resistance. The more the mass in the zone, the more is the resistance and hence there is a higher drafting force. It is well established that this drafting force is inversely proportional to the setting (or nip to nip distance) in each zone. To counteract the higher drafting force in the back zone (owing to more bulk), the setting distance is kept wider in the back zone. Depending on the break draft employed in the back zone, the mass in the middle zone reduces. Accordingly, the margin for the setting over *effective length* is also less in the middle zone. Finally, as the mass in the front zone is much reduced, the margin for setting in this zone is the least amongst the three zones.

As shown below, a simple setting rule was used in earlier days and it became very popular with a conventional 4 over 4 graduated drafting system. It is related to nip-to-nip distance between bottom drafting rollers.

- Front zone nip-to-nip distance = Effective length + 1/8 in (or 3 mm)
- Middle zone nip-to-nip distance = Effective length + 2/8 in (or 6 mm)
- Back or break one nip-to-nip distance = Effective length + 3/8 in (or 9 mm)

The allowances mentioned above may vary from mill to mill and also would depend on the state of material. The allowance may be slightly increased if the fibre arrangement is random. Thus, the settings for the first draw frame passage, following the card, will have to be little wider than those quoted above. This is because the drafting resistance in this case will be comparatively higher. A wider setting thus helps in overcoming this resistance and smoothens the process of pulling or thinning-out. After the first passage of the draw frame, the fibre orientation and their alignment are improved. Correspondingly, the drafting force is also reduced. In such cases, it is not necessary to use wider settings than those quoted above.

8.3 PASSAGE OF COTTON THROUGH CONVENTIONAL DRAW FRAME

As shown in Figure 8.14, the slivers from card cans are guided over a lifter roller which both guides the slivers and lifts them from card cans. Six (or) such slivers are then passed over slotted guide plates, the slots providing the guidance and also separating each of the slivers. In a conventional machine, the slivers are further passed over the stop motion spoons, one each for six slivers. On some other later conventional models, a single preventer roller in place of sliver spoons is positioned

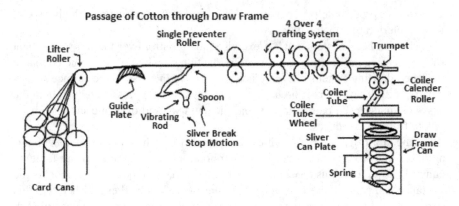

FIGURE 8.14 Passage of cotton through conventional draw frame[1,3]: The passage appears to be the simplest of all machines in the spinning sequence. But fibre control in drafting is of vital importance. This decides the sliver evenness, which governs the short-and long-term irregularities in the yarn.

just prior to the slivers entering the drafting system. In this case, a 9–12-volt electrical supply is given to both the top and bottom rollers. With the material (slivers) passing through the roller nip, the electrical contacts are not made. However, when any of the slivers breaks, the contact between the top and bottom roller is established and the circuit is completed. The electrical relay conveys this to the main supply and stops the machine.

The slivers are drawn and attenuated as they pass through the drafting system. During this, the fibres are parallelized, straightened and aligned to the resultant sliver axis. Each sliver gets individually drafted and emerges in the form of a thin web. The webs of each of the slivers merge into their neighbouring components easily and this helps in their intimate blending. The web across the full width of the head is then collected and simultaneously condensed as it passes through a trumpet. It is finally coiled as a sliver by passing it through a pair of coiler calendar rollers (coiler CR), which are comparatively smaller in size, and through a coiler tube wheel. The coils are laid into a can which is given a small rotational speed to allow the coils to be uniformly spread and distributed.

The stop motion, whether mechanical (in very old machines) or electrical, for detection of sliver breaks, plays an important role. If the stop motion does not act promptly at the time of sliver breaks, the machine continues to run with fewer slivers (may be for a short time). The hank of the sliver produced, in this case, becomes finer than the nominal hank. Also, if there is choking of the sliver in the trumpet or lapping of drafted web around the drafting rollers, it often leads to waste of both material and time. Sometimes it has been found that the stop motions are tampered with by the operators themselves, especially when they find that frequent machine stoppages seriously affect the machine's productivity. As a general practice, the productivity of each draw frame is recorded in the form of hank meters which are fitted on the machine. Frequent stoppages certainly affect the total hanks produced during a shift, for which the operator is usually held solely responsible.

Apart from hank meter, the productivity is also controlled by the hank (its fineness) of the delivered sliver. Depending on the hank organization, the hank of drawn sliver is decided and is adjusted by controlling the total draft in the machine. Ultimately, the level of the total draft used in any drafting system depends on the number of slivers doubled (usually 6/8/10).

$$\text{Total draft} = \frac{\text{Surface Speed of FR}}{\text{Surface Speed of BR}} \text{ and hence,}$$

$$\text{Total draft} = \frac{\pi \times \text{Diameter of FR} \times \text{its rpm}}{\pi \times \text{Diameter of BR} \times \text{its rpm}}$$

$$= \frac{\text{rpm of FR}}{\text{rpm of BR}}$$

(8.1)

(In the conventional graduated drafting system, generally the diameters of the back and the front roller were the same.)

A gearing diagram of a conventional drafting system is shown Figure 8.15. It can be seen that the back roller is driven from the front roller and hence the ratio of the speeds will be proportional to the ratio of gear wheels linking these two rollers.

Assuming that the back roller revolves through one revolution,

$$\text{Number of FR revolutions} = 1 \times \frac{\text{BRW}}{\text{CP}} \times \frac{\text{Crown wheel}}{\text{FRP}}$$

where, BRW refers to back roller wheel, CP to change pinion and FRP to front roller pinion.

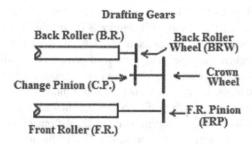

FIGURE 8.15 Gearing of drafting rollers[1,3]: This is very simple and involves two important wheels—BRW and FRP, both of which govern two critical parameters—main draft and break draft.

The expression (A), therefore, becomes

$$\text{Total draft} = \frac{\text{BRW} \times \text{Crown wheel}}{\text{CP} \times \text{FRP}} \tag{B}$$

The change pinion (CP) itself indicates that the wheel can be changed. Normally, except for the CP, all other wheels are expected to be constant. In (B), therefore, omitting the actual value of CP, we get another useful expression

$$\text{Total draft} = \frac{\text{Constant}}{\text{CP}}$$

This constant is called the *draft constant.*

Assuming BRW = 60^T; Crown wheel = 100^T and FRP = 20^T]

$$\text{Total draft} = \frac{60 \times 100}{\text{CP} \times 20} = \frac{300}{\text{CP}} = \frac{\text{Draft constant}}{\text{CP}} \tag{C}$$

As mentioned above, for a given set of wheels, the value 300 remains constant for a given machine. The fineness of the sliver delivered is proportional to the value of the total draft which is varied by changing the value of the change pinion.

Similarly, though the number of teeth on the BRW, crown wheel and FRP are constant for a given set-up, sometimes the need arises to change the draft constant. For example, when the required change pinion is not available, it may be necessary to alter the value of draft constant. This is done by changing the number of teeth on BRW or to that extent, any wheel other than CP (pitch is maintained the same).

Each draw frame machine is split up into what are termed as *heads.* It is that portion of the rollers through which a group of 6 or 8 slivers are made to pass. As a convention, there used to be 2, 4 or 5 heads in a conventional draw frame. On all modern draw frames, there are only 2 heads (or a single head for automatic can-changing machines).

There are usually two draw frame passages used for carded yarns [i.e., Card Sliver → DF → DF →Can-fed Inter/Roving → Ring Frame].

Different arrangements as shown in Figure 8.16 are the outcome of the convenience of placing the draw frames in the mills. This arrangement is for two-delivery machines. With modern machines, a tandem arrangement is more popular as it serves both purposes—minimum transportation and ease of operations for the tenter.

Modern draw frames are driven with individual motors. With such drives, it is possible to plan the layout of the machine better, and utilize the floor space economically and yet spaciously. In the conventional layout, the machines were too crowded which prohibited better material transportation. With a modern setup, the machines with higher productivity use larger cans and long drawn-out creels. Therefore, it becomes necessary that the material transportation system be improved. For this, it is essential to space these machines with adequate back and front alleys. It also

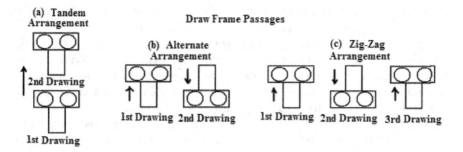

FIGURE 8.16 Placement of draw frames[1,3]: There are, on an average, two passages of draw frames: breaker and finisher - for carded material. They are placed in such a way that the travelling distance for the sliver cans of breaker to reach the creel of the finisher is as small as possible. (a) Tandem arrangement, (b) alternate arrangement and (c) zig-zag arrangement.

allows easy access to the machines and permits better cleanliness around them. It may be mentioned here that in the process of modernization, when fewer modern machines replaced a large number of conventional and slow-speed machines, the availability of space for a neat layout is never a problem. Thus, a properly planned setup helps the tenter to have quick and easy access to mind the machines, especially when he has to attend several of them at a time.

8.3.1 BOTTOM ROLLERS

These rollers, as mentioned earlier, are either made in one continuous length or in sections. When in sections, they are joined together with special joints. As shown in Figure 8.17, projection A on roller X is made to fit into recess B on roller Y (usually called a male–female joint), thus making them one piece of longer length. Normally, the length of the machine is longer with more heads accommodated in one draw frame. A long and continuous bottom roller covering the whole length of machine poses serious problems. When such a long length has to bear the top roller pressure, often the bottom rollers yield to this pressure and have a tendency to bend or buckle.

Another source of trouble is that there is always a chance of their flexing. This is because the rollers are driven from one side only, whereas the other end is merely held in the bearings. Thus, with longer length of bottom rollers, they are always under great

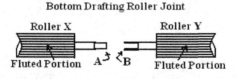

FIGURE 8.17 Roller joints[2,3]: With long continuous bottom rollers, if any portion of the roller is damaged, the whole roller will be required to be replaced. The rollers are therefore made in small sections.

strain due to torque resulting from one sided roller drive. Apart from this, the rollers, at times, may also get damaged at the flutes or worn out at the bearing supports.

Again, if the rollers were of continuous length, it would be necessary to replace the whole roller. However, with joints, only that particular section of the roller is required to be replaced. Almost all modern draw frames, however, have only one or maximum two head and consequently, the length of the roller involved is comparatively much shorter. In addition, the rollers are larger in diameter. In such cases, the rollers can be continuous and normally extends the full width of the machine without any joints.

The grip of the bottom rollers and their power to draw the fibres effectively is very important. The design of the flutes in this case plays an important role. In design, the land of the flute, its height and the angle of the slanting faces of flutes, all decide the gripping power of the flutes. As the bottom rollers have to bear the top roller pressure (loading), their strength is equally important. For this reason, they are made of steel and are often case-hardened. Whereas the portion of the roller supported in the bearings forms the neck, the portion where the slivers run over is made larger in diameter and is fluted. With different makes of machines, the size, the shape and the dimensions of the flutes vary. Usually the number of flutes is related to the roller diameter. For example, 1 in diameter: 36 flutes; 1¼ inch diameter: 45 flutes; and 1½ in diameter: 54 flutes, etc.

Sometimes the arrangement of the flutes is such that they have a varying pitch on the circumference. Generally, with soft covered top rollers and additional weighting on them, the bottom roller flutes make grooves or indentation marks on the softer covers of the top rollers. The cot surface appears to be covered with minute corrugations. This makes the roller surface uneven. The result is that the top soft-cot surface does not hold the cotton material firmly. Normally, when the cot surface becomes uneven, it is a common practice to buff (very light and fine polishing action). But the flute causing the indentation marks is a perpetual problem and frequent buffing is not the solution. With varying pitch of the flutes therefore, *hunting** of the flutes is avoided.

In this connection, it may be mentioned that whenever the machine is stopped for a sufficiently long period of time, it is necessary that the weightings on the top roller should be released; otherwise, the flutes remaining idly in contact with the soft cushioned top rollers definitely cause indentation marks that last forever. This permanently damages the soft cushion cover of top rollers. The length of the fluted portion of the bottom roller is more than what is just necessary to accommodate the total number of slivers fed into the system. In fact, the actual fluted length (Figure 8.18) provided for the roller is a little longer than the actual running width of the slivers. The extra width of the fluted portion is generally provided to permits to-and-fro traversing of the slivers over the major fluted length.

8.3.2 Top Rollers

The necessity of the top rollers being covered by the synthetic cots is evident. If the top rollers were plain, smooth and metallic, the grip between the top and the

---

* Hunting – the same flutes making impressions at the same place on softer top roller]

FIGURE 8.18   Flutes on bottom drafting rollers[2,3]: Whereas the top rollers are cushioned with synthetic rubber, the bottom rollers are made of steel and are fluted. This combination improves the grip.

bottom rollers would never have been satisfactory. Instead, if they were fluted like bottom rollers (Figure 8.19a), with their flutes meshing into each other, the heavy pressure imposed on them would crush the fibres. To avoid this, it would have been necessary to alter their meshing depth (penetration of the top and bottom flutes into each other), whenever the bulk of the material was changed (Figure 8.19b).

Covering the top rollers with soft cushion solves all these problems. Inside the cushion cover, the bare top rollers are solid and are made of steel. In early days, when cork or leather was used as a covering, the bare top metallic rollers were plain or knurled. The synthetic cots are spirally grooved from inside. Similarly, the bare top rollers are also given a light spiralling over their surface. This holds the cot more firmly when it is mounted over the bare inner roller. Much earlier, on conventional draw frames, a soft cushioned solid top roller (Figure 8.20c) was made to work in slides with brass bearings. However, these bearings were good enough for only slower speeds. The solid rollers on these machines were thus used on the last three lines where the speeds were comparatively slow and the friction was low. The use of brass bearings to hold the rollers at the neck was thus restricted to slow speeds only.

However, the front roller runs at a much faster speed and hence is invariably equipped with a loose boss or loose bush type of roller (Figure 8.20a and b). The cushioned length of top rollers, in each case, is made to extend over the fluted portion

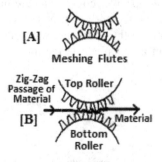

FIGURE 8.19   Meshing flutes[2,3]: To improve upon the grip between the top and bottom fluted rollers, the flutes of both are made to penetrate a little into each other. However, the meshing distance has to be appropriate to suit the bulk of material passing through.

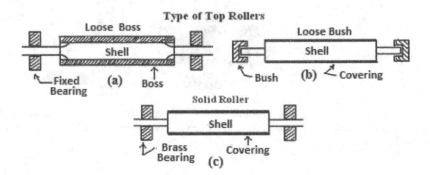

FIGURE 8.20 Loose boss and loose bush rollers[2,3]: With the advent of high-speed draw frames, the front roller speeds were significantly increased. In this situation it was necessary to support these rollers with more precision bearings. The loose boss or loose bush types of front top rollers were essential to cope with the higher speeds.

of the bottom roller in loose boss type, and the centre spindle A (Figure 8.21) has a slight bevelled body over which an outer shell B is fitted. This shell runs loose on the central spindle and makes contact with it mainly in the middle region where the bearings are provided. This reduces the friction considerably.

Lubrication also becomes very easy as the oil has an easy access to reach the central part of the spindle. In the loose bush type system (Figure 8.22), needle bearings are provided in the bush and they help in smooth running of the roller with minimum friction. Yet another advantage in this case is that an oiler is able to oil the bearings when the machine is running. This is because the bushes are well away from the running parts. Also, when the bushes become worn out, only they need to be replaced. When the front roller is *double boss*, it is advisable to use the loose-boss type of rollers, whereas the loose bush type roller should be used when front top roller is *single boss*.

8.3.3 ROLLER COVERING

In earlier draw frames, the top roller coverings were made of cork sheets. These sheets were made by grinding the cork and then mixing the powdered cork with suitable adhesive. The mixture thus obtained was rolled into a sheet and baked. These sheets were then cut and applied on the top rollers like a cloth. The cork

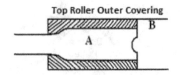

FIGURE 8.21 Loose boss[2,3]: The bearings are provided which hold the outer shell (B) over the inner roller (A). This ensures a very smooth running of the outer shell.

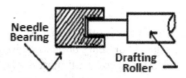

FIGURE 8.22   Loose bush[2,3]: The top roller is covered with soft synthetic rubber. The roller shaft, however, is held by the bush having fine smooth bearings. With very high speeds, the front top rollers are equipped with bush bearings.

cots provided both the cushion and the suitable surface for drafting. An improved method, however, was to have an extra cushioned seamless cork. The cushions were of finely ground cork, extruded through tubular openings to give the correct size of the cots. During fixing of these cots on the bare top rollers, the inner cot surface was glued.

Developments in resilient synthetic compounds have produced several types of roller coverings from an extruded tube of synthetic polymer. When properly cured and cut to the desired length, this type of cover is stretched over the bare top roller and later buffed to a true surface. The material is quite resistant to the lubricating oils used for machinery parts, normal wear and channelling, and has high drafting qualities.

During normal working, the rollers may wear out unevenly or develop hollow places. A light grinding (buffing) is necessary both for a new cot as well as the one in use. The surface irregularities are evened out by buffing. A roller thus can have several buffings during its entire life. However, when the thickness of the cot covering becomes too thin (less than 3–4 mm), the cushioning effect of the cots is severely impaired. The cots at this point must be replaced.

The metallic flutes on both top and bottom rollers are cut much like gears. Their dimensions, however, are much finer. The collars are provided at the end of the boss to keep the meshing of the flutes very precise and controlled (Figure 8.23). This requires a very precise setting. The flutes in this case are made coarser for the third and back rollers, medium for the second, and fine for the front bottom roller. This is in accordance with the weight of the sliver material passing through the rollers. The material passing between the top and bottom metallic fluted rollers curves around the flutes to produce crimp. The crimped length varies according to the depth of the meshing and usually ranges from15–40%. This is often cited as an advantage to get higher production for a given speed of rollers.

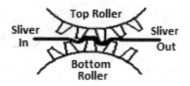

FIGURE 8.23   Bending of sliver[2,3]: The advantages with this type of gripping are that the grip on fibres is far more positive and that, owing to the crimped length, production is increased in the range of 15–40%.

8.3.4 Cots

A good roller covering (cots—Figure 8.24) is expected to give a uniform quality of performance and should be capable of being buffed precisely. It should have properties like anti-eyebrowing, heat resistance, etc. It must not be affected by dyes, textile solvents or lubricants (sometimes lubricating oils are used in very small quantities in artificial fibre mixing). Under certain circumstances, moisture condenses on the surfaces of the top drafting rollers, thus leading to frequent lap-ups. Hence moisture absorbing ingredients are also added to synthetic rubber during manufacturing of cots.

They must not create *start-ups* (when machine is stopped for a long time, the top cushioned rollers are likely to develop this trouble) after long shutdowns.

During their manufacture, the cots are made in varying degrees of hardness, which is expressed as *degree of shore hardness*. The cots with higher shore hardness can withstand higher pressure, and even at higher speeds they do not suffer surface deformation easily.

The softer cots ranging from 60° to 65° R (shore hardness) are used for normal working. Usually the softer cots give notable improvement and lead to better quality of the material processed because of the better grip that they offer. However, they wear-out comparatively quickly owing to their soft nature. In this case, frequent buffing becomes necessary. The cots, which are resistant to lap-ups, contain certain electrolytes. These effectively neutralise the static attraction between the surfaces of the cots and the drafted material. Further, with worn-out bottom steel rollers (flutes),

FIGURE 8.24   Draw frame top roller cots[2,5]: The soft cushioned cots are available with different hardness. Whereas harder cots are preferred for bulky fibres, the softer cots, in general, give better evenness, they wear faster though.

the softer cots perform comparatively better. On the other hand, more resilient fibres (like polyester) require more gripping pressure and hence harder cots (shore hardness 80° to 85°) are required.

With modern high-speed draw frames, the higher loading on the top rollers, along with the higher speed, heats the top roller surface considerably in spite of the use of anti-friction bearings. This is due to friction between the surfaces of top and bottom roller. The conventional glue and cot adhesives, therefore, become unsuitable. In addition to this, if the top roller bearings are not lubricated on a regular basis, the cot surfaces experience a drag, thus resulting in loss of roller revolutions. This dragging further accentuates the tendency for the cots to come off.

Another important feature of high-speed cots is the spiral grooving on their surfaces. These grooves are U- or V-shaped and are moderately deep, to about 1.6 mm (2/32 in). The objective of these grooves is to minimize the lap-up tendencies.

In addition, the fly accumulation on the top surface is reduced. The grooving also prevents surface distortion or bubbling, and helps the rollers to maintain uniform surface at the nip point

8.3.5 COT WEAR AND TEAR

During working, most of the synthetic rubber cots develop channels or hollow portions in the regions where the sliver passes over. This effect is reduced by giving a little traversing motion to all the slivers passing through the roller nips. This ensures even and uniform wear of the cot surface by spreading the effect over a larger width.

As shown (Figure 8.25), an eccentric gives oscillating motion to the bell crank lever fulcrumed at f, and this in turn makes the pawl push the ratchet. A small worm on the ratchet shaft drives the worm wheel on another shaft, A.

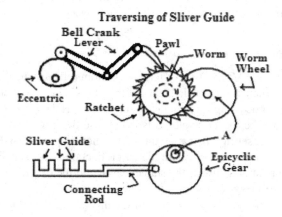

FIGURE 8.25 Traverse mechanism[2,3]: In the absence of this motion, the slivers would repeatedly pass through the same portion of drafting rollers. The top roller cots being softer, this would seriously wear out only that portion of the softer cushion, thus resulting in channelling.

There are two epi-cyclic gears mounted on A and they give the to-and-fro motion to the sliver guides.

The channelling has been attributed to some form of creep or plastic flow. It appears rapidly during the early life of the cot and thereafter increases very little. This can be corrected, as stated, by light grinding (or buffing). After each subsequent grinding, the surface shore hardness increases, resulting in loss of cushioning property of the cots. Many cheaper roller coverings are available and give good initial performance. However, at higher load requirements, their useful life is much less and they quickly deform when the machine stops.

8.3.6 Maintenance of Cots and Current Requirements

It must be borne in mind that not all the lap-ups can be attributed to static problems. They are also caused by the attraction between the fibres and the surface of the cots owing to presence of dirt, moisture, oil and grease. For ordinary cleaning purpose, it is sufficient to use a rag dipped in mild detergent to wipe away these from the surface of cots, followed by rinsing in clean water. In the case of more stubborn stains, carbon tetrachloride can be used, provided suitable precautions are taken against the toxic fumes that are produced during the operation.

There are other restricting conditions. With the recent trend for higher speeds, operating conditions have become increasingly arduous in terms of coping with new types of fibres, higher weightings, and increased air flow for air conditioning and pneumatic clearers. The dimensional accuracy called for the cots, in this case, is in the range of 0.025 mm (approx. 1/1000 in).

The use of precision fluid seal bearings has become necessary. The presence of lubricants on the fibres and on machine parts requires a fair degree of oil resistance. The resistance to ozone cracking is another requirement, and this is due to constant air flow around the roller surfaces and increased air changes. The ozone cracking effect can be much reduced by the use of a certain type of synthetic rubber; however, the resistance to fibre adhesion of the cots deteriorates.

8.3.7 Diameter of Rollers and Roller Settings

With the conventional 4/4 drafting system, the bottom roller diameters (Figure 8.26) were mostly governed by the staple length of the fibres. Thus, smaller diameter rollers were used for processing short staple cottons, whereas the larger diameter rollers were preferred for the longer staple. This is because the larger diameter rollers do not permit the closer settings that are required for short staple cotton. However, using the larger diameter rollers does permit higher production rates. The Table 8.26 gives an approximate relation between the staple length and the roller diameter used in a conventional drafting system.

While setting the rollers in the conventional draw frame, a broad principle is used as a guideline. The distance between the nipping points of succeeding pairs of rollers (taken from front to back) goes on increasing. The first pair of the rollers is thus at distance D_1 (Figure 8.27) which is just greater by approximately 3 mm than the staple length of the cotton. The middle roller setting D_2 is further

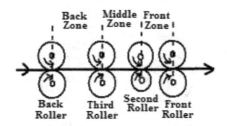

FIGURE 8.26 Drafting rollers[1,3]: With a 4/4 conventional drafting system, the roller speeds are gradually increased from back to front roller. So also, the roller settings are gradually reduced from back to front zone. The typical diameters of the bottom rollers used in relation to the type of cotton (staple length) are given in Table 8.1.

increased by 3 mm. The back setting D_3 is the widest and is still further increased by 3 mm.

The plan holds good. However, when there is greater variation in the staple, a slight reshuffling is required, especially when the cotton is softer. It may be also noted that the second roller is invariably of smaller diameter so as to facilitate a closer setting in the front zone. The following example will give an idea about the roller settings and the gauges to be used in the three drafting zones (Figure 8.28a).

Condition No.1: Staple length processed = 26 mm

Setting = Centre to centre distance between F.R. & S.R.: (A_1)

A_1= staple length + 3 mm **

[** 3 mm is an allowance given over the staple length]

= 26+3 = 29 mm

The rollers are spaced at distance A_1, which is the setting and is equal to staple length plus an allowance (29 mm). However, the rollers are actually set by putting a plate of the required thickness between the two bottom rollers. Hence, to find the gauge of the plate (G_1 in Figure 8.28b), it is necessary to subtract a value of half the diameter of the rollers involved from the setting A_1.

TABLE 8.1

Conventional Draw Frame Rollers

Cotton & Staple	Drafting Roller Diameters in mm				Position
	FR	SR	TR	BR	
Indian cotton up	24	24	24	24	Top
to 26 mm	26	23	26	26	Bottom
American cotton	26	26	26	26	Top
up to 30 mm	29	26	29	29	Bottom
Egyptian cotton	29	29	29	29	Top
up to 37 mm	32	29	32	32	Bottom
Sea Island cotton	34	34	34	34	Top
up to 42 mm	37	34	37	37	Bottom

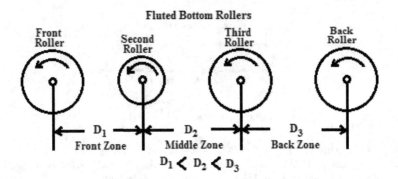

FIGURE 8.27 Bottom roller diameters and their settings[1,3]: The setting depends on the diameter of the rollers and the staple length processed. Some allowance over and above the staple length is necessary to decide the nip-to-nip distance.

Thus,

$A_1 - (d_1 + d_2)/2 = G_1$ (roller gauge in mm in the front zone): (X)

Taking the values from Figure 8.28b and the value of A_1

$G_1 = 29 - [29/2 + 26/2] = 1.5$ mm ---Roller Gauge in mm in front zone

Condition No. 2: Staple length processed = 24.5 mm

Here staple length + allowance = 24.5 + 3 = 27.5 mm

Using the same format, we will get

Roller gauge in mm = 27.5 − [29/2 + 26/2]

= 27.5 − [14.7 + 13.0]

= 27.5 − 27.5 = zero

The roller gauge of zero means that the rollers are made to almost touch each other. However, in actual practice, the two surfaces are set at the thinnest possible distance, equal to the thickness of a piece of paper. Hence the setting is commonly

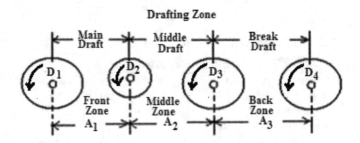

FIGURE 8.28a Bottom roller diameters and their setting[1,3]: Except for the second bottom roller, all other bottom rollers are comparatively large in diameter. The second roller is smallest among all other rollers and facilitates closer setting in the front zone—the main drafting zone.

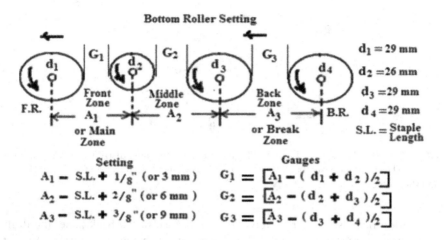

Bottom Roller Setting

$d_1 = 29$ mm
$d_2 = 26$ mm
$d_3 = 29$ mm
$d_4 = 29$ mm
S.L. = Staple Length

Setting	Gauges
$A_1 = $ S.L. $+$ $1/8''$ (or 3 mm)	$G_1 = [A_1 - (d_1 + d_2)/2]$
$A_2 = $ S.L. $+$ $2/8''$ (or 6 mm)	$G_2 = [A_2 - (d_2 + d_3)/2]$
$A_3 = $ S.L. $+$ $3/8''$ (or 9 mm)	$G_3 = [A_3 - (d_3 + d_4)/2]$

FIGURE 8.28b Roller gauging[1,3]: The distance between the any two adjacent rollers is actually set by inserting appropriate rectangular plates of specific thicknesses in between the two rollers. Finally, the thickness of the plate itself (the gauge) is arrived at considering staple length, allowance over the staple length and the diameters of the rollers involved.

referred to as *paper gauge*. This small distance (paper gauge) ensures that the two rollers involved in setting do not touch each other, but are still able to fairly maintain the closeness as per the setting requirements.

Further, it is also obvious that had the staple length been less than 24.5 mm, the required roller gauge would have been even less than zero—a negative setting, which is impossible. The need for the smaller diameters of the rollers for shorter lengths thus becomes evident.

8.3.8 ROLLER SETTING IN RELATION TO DRAFTING AND BULK

As mentioned earlier, in a graduated drafting system, each succeeding pair in the drafting system moves progressively faster. When the fibres enter this system, they are gradually subjected to an accelerated motion. The allowance of 3 mm given in the front zone, in many cases, is adequate enough. It will be obvious that this allowance corresponds to a distance BP in Figure 8.29 and may vary on either side, depending on the class of cotton that is processed. The fibres under this condition, therefore, yield to the action of the rollers progressively running faster. In doing so, their contacts with those around, set up sufficient friction to cause their straightening. The surrounding fibres are also acted upon in the same manner. The faster moving roller pairs pickup comparatively fewer fibres from the preceding pairs and this results in a thinning-out of the material.

The settings between the middle zone and the back zone are further increased so that the former is wider by 6 mm and the latter by 9 mm. Thus, whereas

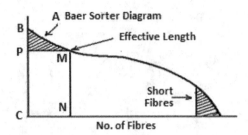

FIGURE 8.29 Baer sorter diagram[2,3]: It gives the complete fibre distribution of the lot processed. The two important fibre properties revealed through this diagram are effective length and short fibre percentage.

the setting increases progressively from front zone to the back zone, the draft reduces from front to back zone. The reason why each zone is not uniformly set apart is not difficult to understand. The bulk of the material in the back zone is maximum among the three zones. Therefore, the resistance, in terms of fibre-to-fibre friction offered against the pulling force of third roller, is also at the maximum level. By setting the rollers at a wider distance, it enables the fibre bulk to get draft easily. Further, the higher friction between the fibres in this back zone imposes higher strain on the fibres. The widening of the setting in the back zone greatly relieves this strain and helps easy attenuation. Thus, it avoids the possibility of any likely fibre damage. It may be mentioned here that, if the top rollers are not adequately loaded, the heavy bulk of the material in this zone may lead to the top rollers slip.

The bulk is reduced in proportion to the draft employed in the back zone, and hence the setting in the middle zone is slightly narrowed. The setting in the front zone, where the draft employed is at maximum level, is the narrowest of all the zonal settings. This is because the thickness of the material in the front zone is still further reduced. There is another reason for this narrow setting in the front zone. As the draft is much higher, the fibres experience a sudden increase in the acceleration in this zone. The time taken by the fibres to travel in this zone is thus much less. This necessitates much better control over the fibre movement. It therefore requires a closer setting in this zone. Even then, as explained before, the allowance of 3 mm in this zone is still able to protect the fibres longer than effective length. As a general rule, the roller setting in any zone is inversely proportional to the bulk processed in that zone. A similar rule can also be deduced in regard to the draft and the setting. Thus, the draft employed in any zone is always inversely proportional to the roller setting, the higher draft requiring closer setting and vice-versa.

If the fibre length of the cotton is irregular (wide scattering of the fibre length distribution), it is better to bring out the finer sliver. Whenever any two rollers are set wide apart, only a small tension draft is employed between them. When the speeds are higher along with higher draft, too much fly and fluff is generated, and this is why the high-speed draw frames must have suction hoods over the drafting system.

8.3.9 ROLLER WEIGHTING

In conventional draw frame, the rollers were usually weighted by hanging the dead weights. The inks with the hooks around the necks of the rollers were placed on either side (Figure 8.30a). The common system of roller loading with this type weighting system was: FR - 22 lb; SR–20 lb; TR and BR–18 lb each on either side. With the weights on either side, the actual load was double these weights.

Traditionally, the front roller was loaded with the most weight, as the main drafting operation was carried in the front zone. In some cases, the actual weightings were changed depending on the class of cotton; heavier weights were used for lower grades. Even then, the hanging weights, owing to the machine vibrations, used to dance vertically.

This seriously affected loading on the top rollers and whole system became incapable of providing consistent loading conditions. Around 1960, M/s Whitin (Figure 8.30b) introduced their high production draw frame with top arm loading. The powerful springs enclosed in hollow metal cylinders carried strong, synthetic, sturdy bushes at their lower bottoms which were made to sit on the bush bearings carrying the roller necks. The loading system was unique because it was free from the dancing effect of the weights, and offered much higher and more precise pressure on the top rollers

8.3.9.1 Weight Relieving Arrangement

In any type of roller weighting system, the weight relieving motion is a must. This is because with the pressure acting on the top rollers it would be difficult to remove them at the time of lap-ups or cleaning. On conventional machines, a simple eccentric connected with a handle facilitated easy raising of the weights. When the weights were raised, the pressure on the hooks around the necks was released. The hooks were then taken off and the top rollers were removed for cleaning or any other purpose. The top rollers may also need to be removed at the time of roller buffing.

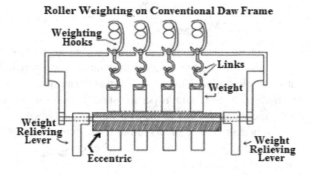

FIGURE 8.30a Conventional roller weighting[1,3]: In older conventional roller weighting systems, the rollers were weighted using hooks-links and hanging plate weights. Jumping of these weights was caused by the vibrations set in the machine during its working. This caused large variations in roller loading leading to roller slip.

FIGURE 8.30b   Top arm weighting: This led to revolutionary development in roller weighting system. With the top arm and spring weighting, the precised pressure on top roller has been made possible. The compressed spring loading has thus significantly reduced the roller slip. (Photograph showing top arm weighting on conventional draw frame.)

The weight relieving system has another important role to play. As mentioned before, when the machine is stopped for a longer duration of time, for major repairs or for any other reason, it becomes essential to relieve the top roller loading. This is because when the rollers are loaded for some time in idle condition, the bottom roller flutes leave long-lasting indentation marks on the softer cots of the top roller, thus permanently damaging their surfaces. In a modern top arm weighting system, the top arms, through a simple raising arrangement, can be lifted up, or in the case of pneumatic loading, the pressure can be simply shut off. This relieves the top roller loading and avoids any possible damage to the cots.

8.4   FUNCTIONS OF DIFFERENT STOP MOTIONS ON DRAW FRAME

The basic function of any stop motion is to stop the machine when any of its important mechanisms fail or any of its technological requirements fall short in their performance. The front and back stop motions mainly see that the number of slivers fed to the system and that delivered by it remain as per the specification. In conventional draw frame, for example, six slivers are fed to the drafting system, whereas a single sliver strand is delivered. When any of the slivers fed to the system breaks or the supply of any one of the feeding cans gets exhausted, the number of slivers entering the system reduces. This changes the hank of the sliver delivered. The machine must stop at this time to enable an operator to correct the fault. Similarly, when a thicker sliver passes through the delivery end, it invariably chokes the coiler trumpet. This may result in a huge amount of sliver accumulating in the region between the front roller and the coiler, thus producing large amounts of *soft waste*. The machine must stop again. Even in the case of roller lapping, its early detection by the stop motion can control the soft waste and prevent a defective sliver from being coiled into the can. Though good sliver waste can be reused, it often leads to processing difficulties and increases irregularities in the delivered sliver. The time lost in correcting the defect, when not noticed in-time, is yet another factor. If the stop motion does not detect these faults owing to their malfunctioning, or if there is no stop motion at all, to provide for such acts, then it may either produce faulty material or would lead to serious damage to the machine parts. In addition, the worker will be under constant anxiety.

Apart from producing soft waste, sliver missing at the back creel, or even roller lapping, eventually lead to a comparatively thinner sliver delivered. This again leads to variation in hank of material delivered. The hank variation seriously affects the quality of the sliver delivered. It is known that this variation in the hank contributes to the within-bobbin variation in the final yarn and increases overall CV% of the yarn count.

A very old method of detecting a sliver break mechanically and then acting upon it to stop the machine is shown in Figure 8.31. In the normal position, when the sliver passes over the spoon, it presses the top of its portion around the fulcrum F. This causes the notch at the bottom of the spoon to remain out of path of the vibrating arm. When the sliver breaks, the notch falls down and comes in the way of vibrating arm. The eccentric, in the normal position, provides an oscillating motion to the rocking shaft. When the sliver breaks, the notch comes in the way of vibrating arm. However, the motion from the eccentric leads to lifting of the lever A, and this results in knocking off the starting handle. Finally, the belt is shifted on the loose pulley to stop the machine.

With single preventer rollers D and E, electrical connections are made with both the top and the bottom rollers (Figure 8.32). As long as the material passes between them, the electrical circuit remains disconnected. When the sliver breaks, the bare surface of D and E touch each other and complete the circuit to energize a magnet that activates the belt-shifting mechanism. It must be mentioned here that the cotton, when dry, does not conduct electricity. However, with a higher level of humidity, the moisture-regain level partially allows the flow of electrical charges through the

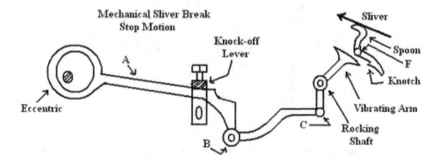

FIGURE 8.31 Mechanism of back-stop motion on conventional draw frame[1,3]: It is partly mechanical and partly electrical stop motion. The break is sensed mechanically but the signal to stop the machine is sent electrically. Mechanical detection of sliver break.

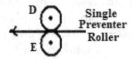

FIGURE 8.32   Single preventer roller[1,3]: The machine runs as long as the electrical contact remains broken by the slivers passing through the nip of the rollers. A broken sliver re-establishes the contact and stops the machine.

strand. In the rainy season, especially, this is further aggravated and hence there is frequent malfunctioning of this type of stop motion, even when the sliver may not have actually broken.

A good stop motion working precisely can always be an asset to a tenter and is beneficial for both controlling soft waste and maintaining the regularity of the sliver. Faulty indications of sliver break or jamming at trumpet, however creates a nuisance for the operators who tamper with the electrical connections.

Further, the detection of sliver break and subsequent stopping of the machine should be quick. The machine must stop before the material reaches the drafting system. This is very essential for piecing of broken sliver. The tenter otherwise will have to take a fresh sliver end through the drafting system—an act which is not only time consuming but rarely practiced. Instead, what the tenter normally does is lay the fresh sliver end over and across the remaining slivers on the creel table (in modern draw frames) and allow it to be taken by the other sliver to the drafting system. The missing portion, which has already entered the drafting system, however, is finer in hank and this leads to hank variation in the delivered sliver. On modern machines, when the sliver breaks, the relaying is directly connected to the motor which then stops. In such a case, the brakes are applied instantaneously to stop the machine.

In conventional draw frame, at one time the full can stop motion worked on the upward pressure developed by the material in the can. The sensing and the actual execution were then mechanical. Modern machines have meter counter on which it is possible to set a certain length (Figure 8.33).

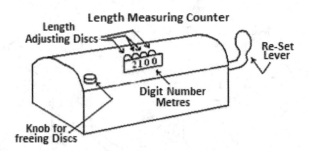

FIGURE 8.33   Stop motion counter[1,3]: Earlier, the sliver coiled in the can was made to overflow its normal height. The heap of this sliver often used to get disturbed during transportation. In many cases, the workers had a tendency to tuck-in the heap from one side into the sides of the cans. The sliver coils at the top thus got disturbed. With digital counters, the precise length is coiled into the can to avoid any overflowing.

When a new can is started, the counter is reset to read the desired length which needs to be coiled. Subsequently, when the machine is run, the counter starts reading the number of meters coiled into the can in reverse order. When it finally reaches zero, the machine stops and activates the mechanism for the change of the can. With automatic can changing mechanisms, a fresh empty can is automatically transferred into coiling position and the machine again starts delivering the material.

8.5 ROLLER CLEARERS

The purpose of the top and bottom clearers is to clean the rollers and collect the short, broken fibres and the dust picked-up by the drafting rollers. This improves the ability of the rollers to draft well and avoids basic problem of lap-ups. In a conventional draw frame, the top clearer was an endless belt made of flannel and was run intermittently at a slow speed over the drafting rollers. A scraper blade A working on the flannel (Figure 8.34) and having to-and-fro motion, used to scrape off the impurities adhered to the flannel.

The gathered mass behind the scraper used to be periodically removed by the tenter. The bottom clearers were usually wooden pieces cushioned from inside by the flannel and were made to contact the bottom steel fluted rollers from the underside.

The accumulated dirt and the fluff from bottom clearers were also periodically removed by the tenter. The whole operation, though minor, is very important as there is always a chance of this accumulated mass following the path of the sliver. If the mass, thus picked-up, is added at regular intervals to the flowing material, it would either add irregularity in the sliver or, in extreme cases, would choke the trumpet. Especially with the latter, this used to lead to heavy waste. This was because, with mechanical or electrical stop motion on the conventional draw frames, the choking at the trumpet did not register a sliver missing and therefore

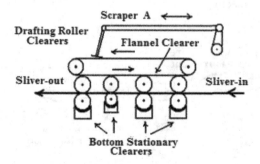

FIGURE 8.34   Top and bottom clearers[1,3]: In processing cotton fibres, the rollers often become sticky due to waxy content on the surface of the fibres. They can also become sticky for other reasons. This develops a tendency for the dirt and short fluffy mass to adhere to the roller surface. The provision of the roller clearer becomes necessary in such cases.

did not sense a break. As there was no sensing of break or missing sliver, there was no relaying of a break. The machine continued to work producing a large amount of soft sliver waste behind the trumpet.

8.6 TRUMPET AND COILING

The function of the trumpet is to condense the web coming out from front roller nip into a compact sliver. The aperture or the bore T (Figure 8.35a) is chosen according to the sliver thickness so that it allows the sliver to pass through easily.

A large bore may allow the free movement of the sliver. But it will not condense the sliver adequately to make it more compact. When the sliver is more fluffy (not compact), it occupies more space and it is not possible to coil a longer length of the resultant bulkier sliver into the can. The compactness of the sliver also helps in avoiding its stretching in the subsequent process. A proper condensation thus helps and is measured by a tapered pin gauge G (Figure 8.35a) marked along its length. The following relation can be used as an approximate guide for choosing the bore size.

$$\text{Diameter in inches} = C \sqrt{(\text{weight of the sliver in grs/yard})}$$

Here, approximately, C ranges from 0.20–0.22 for card sliver, 0.17–0.19 for first passage drawing sliver and = 0.16–0.18 for second passage drawing sliver. The distance L between the aperture and the bite of the coiler calendar roller (Figure 8.35b) should be a little more than the staple length. This is because these rollers also exert a slight pulling action on the sliver coming through the aperture and this is likely to cause a little strain the fibres. The distance (L), little wider than the staple length allows some sort of relief to the passing fibres. The correct size of the trumpet can also act as a *lump breaker*. It is found that when the smallest workable size of the trumpet is used, it improves the sliver evenness. In this condition, the trumpet is able to exercise some control over the passing material and smoothens-out smaller strokes down the projecting fibres.

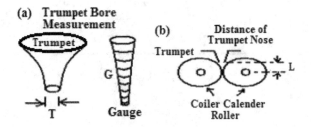

FIGURE 8.35 (a) Trumpet and (b) coiler calendar roller (CR)[3]: The correct size of the bore of the trumpet is directly related to the mass of the sliver passing through it. So also, the distance from the nip of the calendar roller to the bore is very important and is interconnected with staple length processed.

8.7 LIMITATIONS OF GRADUATED DRAFTING SYSTEM

All the natural fibres vary in their length. However, when the roller setting is done, it is based on the staple length or effective length. The allowance given on and over this length usually takes care of longer fibres. In a graduated drafting system, as the pairs of top and bottom rollers are positioned in a vertical line (line of joining their centres is vertical), the distance between their nipping lines becomes far greater for shorter fibres. Ideally speaking, any fibre which leaves the nip of a slow moving back pair must be immediately caught and gripped by the faster moving front pair—a phenomenon which rarely occurs with short fibres. These fibres, for a short duration, travel without the control of either of the pairs, and during this period, their movement through the drafting zone becomes unpredictable.

8.7.1 DRAFTING WAVE

Consider the two pairs of rollers, A and B (Figure 8.36), the former running faster than the latter. At any time, during the movement of the fibres through the system, some fibres will be under the nip of A (fibre D); some fibres will be under the nip of B (fibre E), whereas the remaining shorter fibres (C) will be merely floating between the two pairs. The fibres D and E will precisely move at the speed of respective pairs A and B. However, fibres C having just left B, are not yet gripped by A. The fate of these short fibres (C) ultimately depends on their positions relative to the neighbouring fibres and the influence that they will have on fibres C. If the fibres D have more influence on fibres C, then the latter will be pulled ahead prematurely and out of their turn. Thus, they would not be allowed to have their natural movement through the drafting zone.

In comparison, when fibres E have more influence on fibres C, the latter would be dragged behind and again not allowed to have their natural movement through the drafting system. In both the cases, the action between the pair A and pair B is most

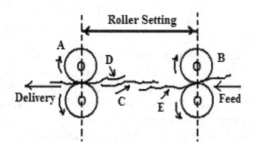

FIGURE 8.36   Floating fibres[2,3]: Normally the ideal drafting implies that as the fibres are left by the nip of the back pair of rollers, they should be immediately controlled by the succeeding pair. Owing to the typical nature of fibre distribution, the fibres shorter than the setting do not follow this. Depending upon how short they are, they correspondingly remain uncontrolled for a short time in the *drafting process*. This leads to a phenomenon called '*floating fibres*', which ultimately produces a typical defect in the product – '*drafting wave*'.

likely to misplace these fibres (C) during the drafting. When the fibres C are pulled ahead, out of their turn, they are eventually in excess at some point and, produce a thick place, whereas when they lag behind, they are missed at some other place which creates a thin place. If the fibres are shorter in length, they remain uncontrolled for a longer duration and consequently the phenomenon of thick and thin places becomes more predominant. These short floating fibres pose many problems in a graduated drafting system with a vertical line roller arrangement.

8.7.2   ROLLER SLIP[3]

The success of the drafting operation lies in precisely thinning out the material. The bottom rollers, though positively driven, cannot carry the operation alone. For pulling the fibre strand through a drafting system, it is essential that they are gripped firmly by a pair of both bottom and top drafting rollers. As mentioned earlier, the top rollers are weighted for this very purpose. However, the top rollers depend on bottom rollers for their drive and it is done by the surface contact between the two.

This arrangement would have been satisfactory if there was no material passing through their grip line. With the slivers passing between the top and the bottom roller, the contact between them becomes indirect. Obviously, with the greater bulk of material, the contact becomes still more indirect. This leads to *roller slip*. Actually, in this case, the surface speed of the top rollers, at times, is either slightly higher or lower than its corresponding bottom roller. As synthetic fibres are more resilient, there is higher slippage of the top rollers. Naturally, the roller weighting in this case has to be increased.

Even when a due allowance is given for protecting the longer fibres, in some instances some of the fibres in the mixing are still longer than the actual roller setting (centre to centre distance). In such instances, it is likely that both ends of these longer fibres are gripped simultaneously. In this case, more powerful pulling action of the leading pair can either break the longer fibres or else simply pull the longer fibre ahead. In doing so, they cause the top roller of the back pair to slip ahead. In some situations, the reverse may also occur and the back pair holding the longer fibres more firmly may force the top roller of the front pair to slip in the backward direction. This again is *roller slip*.

The amount of draft between the two pairs of rollers is also an important factor in this case. The drafting involves a relative movement of the fibres over each other. The sliding friction among the fibres resists this movement. The drafting force has to overcome this frictional force. This is provided by the fibre tension during drafting and is also owing to firm nip between the top and bottom roller. The drafting force (a force required to pull apart the fibres) is proportional to the number of fibres in the drafting zone and hence it is inversely proportional to the draft in the zone. The drafting force increases as the roller setting is narrowed down. Its magnitude also depends on the nature of the material, degree of entanglement and fibre parallelization. All these factors affect roller slip.

Roller eccentricity and cot hardness play their role in modifying a roller slip condition. The greater eccentricity often leads to more tendencies for the rollers to slip.

The higher drafting speed further aggravates the situation. For a certain speed, if the roller eccentricity is within tolerance, when the speeds are increased, it accentuates the effect. Hence the tolerance, in such a situation, has to be narrowed. As mentioned earlier, the softer cots yield better grip and thus reduce the slippage. However, when the need for the harder cots arises, it is essential to use higher roller loading to control slip.

The weighting on the top roller directly controls the tendency for roller slip. This is because it is the proper and adequate loading that establishes the effective nipping between the top and bottom rollers. However, the roller weighting cannot be indefinitely increased due to mechanical limitations. In modern draw frames, therefore, the rollers are made sufficiently stronger by making their diameters larger and using improved metallurgy.

As mentioned earlier, the drafting force which has to overcome the frictional resistance of the fibres being drafted depends on the fibre arrangement in the material processed. The fibre arrangement in the card sliver is most random and hence the drafting force is much higher. With each successive drawing operation, the fibres become better oriented and more parallelized. Accordingly, the fibre extent is progressively improved. Therefore, the effect of roller slip is more pronounced in the early stages of drafting operations. Besides, with the graduated drafting system, the second and third top rollers (especially the second roller) have more tendencies for roller slip.

The roller slip leads to a periodic variation of 8–10 in wavelength (the length over which the irregularity peaks appear at a regular interval of 8–10 in Figure 8.37). In addition to this, there is a tendency for the drafting wave to come into phase at second draw frame passage when processing card sliver and piecing wave to come into phasing at the post comb draw frame. The phasing brings the respective thick or thin places together when processing a group of slivers. More are the drafting regions in

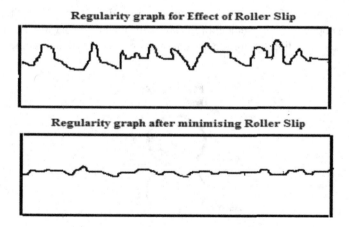

FIGURE 8.37   Periodic variations due to Roller Slip[1]: It is possible to take some measures such as improving the roller pressure, revising the draft distribution or modifying flute design to minimise (or totally eliminate) the effect of roller slip.

any drafting system; more is the tendency for phasing. Therefore, the sliver evenness is not appreciably improved as expected from the process of doubling. In fact, in some instances, the material appears to be slightly more uneven after passing through draw frame.

The roller slip can be best avoided by controlling the unbalanced drafting forces which are experienced by the fibres in front and second zones. Also, owing to the fact that the slivers at different stages have a different order of fibre arrangement, the same machine conditions may not be ideal for giving good results at different stages of the drafting operation. The machine parameters must, therefore, be changed appropriately.

8.7.3 RELATED MECHANICS OF ROLLER SLIP[3]

Even when the drafting rollers are correctly weighted and maintained, roller slip can still occur. In a 4/4 drafting system, with a draft of six, the draft distribution in conventional draw frame is as follows:

Back zone – 1.10 to 1.30

Middle zone – 1.70 to 1.85

Front zone – 2.70 to 2.90

The rollers which are prone to slippage are the second and the third top rollers. Considering the sliver passing through two rollers, A and B (Figure 8.38), when the top roller rotates, an equal and opposite force, F_1, is present on the sliver. This is similar to a force of friction which is generated when there is a motion given to any body.

There is another force, F_3, which compresses and straightens the fibres. Both these forces oppose the movement of the sliver through the roller nip. But the sliver has to move in the direction of the motion of the rollers. Let this force be F_2. Thus,

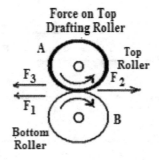

FIGURE 8.38 Drafting force[2,4]: During drafting, the opposing force experienced by the fibres being pulled is the drafting force. It is governed by the setting in the zone, the fibre mass involved and the fibre characteristics.

there are forces acting on either side of the pair of rollers. The rollers, however, must nip the material more firmly than the opposing drafting force to avoid roller slippage.

Even within a sliver, the top surface travels more slowly than the one in contact with the bottom roller. Equally possible is that, in certain circumstances (roller slipping ahead), the top roller can run faster than bottom roller. This may be due to frictional contact and drafting force. Thus, the sliver speed, on many occasions, is equal neither to the bottom roller nor to the top roller. It is governed by tensile strain and the forces involved in compressing the sliver. When the rollers are loaded, the top roller covering is compressed. This leads to deformation of its surface. This is the reason for the speed variation in sliver delivery.

At times, the value of F_2 may be very close to the combined force of F_1 and F_3. In this condition, there are unbalanced forces acting on the top roller. Its speed starts varying between the speeds of back and front line rollers. Thus, the roller may slip in the forward or backward direction. In addition, machine vibrations also influence the contact between the top and bottom roller. Thus, when the rollers vibrate, contact between the top and the bottom roller is constantly disturbed.

8.8   SHIRLEY DRAFT DISTRIBUTION[1]

The Shirley Institute (presently The Textile Institute), Manchester, had suggested some changes in the graduated drafting system so as to reduce the roller slip.

1. **Modified Fluting:** Shirley had redesigned special bottom roller flutings (Figure 8.39a and b—enlarged view of flutings). The flutes were narrower (smaller angle) and were made deeper. This gave more gripping action. The pitch and the land of the flutes were maintained the same, so that the number of flutes and their denting effect on the top roller cover was not altered.

   The narrower flutes and their greater depth cause the top roller cushion to bend a little around the flutes more conveniently and conclusively to exercise better grip.

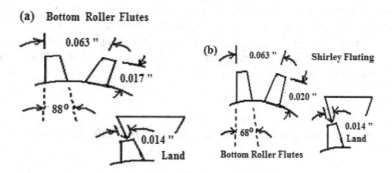

FIGURE 8.39   (a) Ordinary flutings and (b) Shirley flutings[1,3]: The design of the flute is very important as it decides the grip between top and bottom rollers. So also the pitch, the depth, the land and the taper angle of the tooth—all influence the grip. Shirley modified the flute design to improve the grip to overcome the problem of roller slip.

2. **Modified Draft Distribution:** The graduated drafting system gives a progressive increase in the draft from back zone to front zone and this, as discussed earlier, led to unbalanced forces on second and third top roller. Shirley suggested modified draft distribution.

   It can be seen from Table 8.2 that, in Shirley, the middle zone was kept as a neutral zone, separating the back or break zone and the front zone. The unbalanced forces on the second and third rollers were thus greatly reduced. The higher break draft in the modified Shirley distribution was more or less equivalent to a combined draft in earlier graduated distribution. Thus, the bulk of the material in the main zone remained almost the same.

3. **Modified Roller Setting:** As there was just a nominal or tension draft in the middle zone, the roller setting in this zone was made much closer, sometimes even within the staple length to offer a double-grip action on the fibres. The higher break draft resulted in more effective breaking up of the heavy mass fed. This allowed a comparatively closer setting than that followed in that zone in graduated draft distribution. The roller setting in the main front zone was generally not altered.

4. **Modified Roller Weighting:** To exercise better grip, the roller weightings as suggested by Shirley were increased. Thus, the weights on the second and third top rollers were especially increased to 27 kg (60 lb).

Even with the Shirley system, the carded sliver in the second passage of draw frame showed some deterioration in evenness as compared to that in the first passage. This was, as mentioned earlier, due to a phenomenon called *phasing*, where the drafting waves produced at first passage came together and enhanced unevenness. However, the reduced middle zone setting and the draft, as suggested by Shirley, helped in minimizing this defect. The front roller weighting was also increased up to 45 kg (100 lb).

In a Shirley draft distribution, the reduction in the number of drafting zones from three to two very positively helped in reducing phasing. This is because it has been found that with higher number of drafting zones, there are more tendencies for phasing. In all modern draw frames, this principle is adopted. Thus, appropriate changes are made in four-roller drafting systems where the draft in the back zone is of the order of 1.8 to 2.0 and the remaining draft is put in the main zone. The middle zone is a neutral zone and helps in keeping the two zones separate. Ultimately, all these improvements helped to reduce roller slip.

TABLE 8.2

Comparison of Graduated and Shirley Draft Distributions

Draft Zone	Graduated Distribution	Modified Shirley Draft Distribution
Back Zone	1.25	2.0
Middle Zone	1.8	1.03–1.05
Front Zone (Main)	2.66	2.95

8.9   BULK PROCESSED

Every draw frame drafting has a limit up to which it can process the bulk of the material. It generally depends on the effective gripping that the top and the bottom rollers together can exercise on the material being processed. This, in turn, depends on the roller flutings, weightings, roller coverings, amount of draft and the production rate. Even the type of fibres processed can influence this limit. Some of the machinery manufacturers, therefore, always recommend the limit for the amount of material that should enter the main drafting zone. The value of the break draft and hence the back zone setting may be correspondingly adjusted to suit these requirements. Thus, depending on the roller weighting, when a certain break draft value is chosen, the bulk of the material entering the main drafting zone gets automatically adjusted and does not exceed the maximum set limit.

8.10   TENSION DRAFT

The slivers emerging from the front roller are in the form of a semi-transparent web. The web collected in the front is delicate and as such should not be stretched. The level of the draft in this region should be as low as possible. Usually a mere tension draft is applied in this region so that it keeps the web straight and does not allow it to get folded or wrinkled. Another reason for controlling the amount of draft here is that the distance between the front roller nip and the coiler calendar roller nip is quite wide, and as such, large variations are likely to occur if the draft is not controlled. A basic principle, as mentioned earlier, is that when the distances are wide, the draft needs to be just as small as possible. Normally the values of the tension drafts chosen range between 1.03 and 1.05

Short staple cottons are coarser, so when they are stretched in drafting and then released, they often tend to spring back (shrink). For such cottons, the apparent tension draft is still higher. This necessitates a further reduction in the level of tension draft mentioned above.

REFERENCES

1. Manual of Cotton Spinning – "Draw Frames, Comber & Speed Frames" – Frank Charnley, Textile Institute, Butterworth Publication, 1964, Manchester
2. A Practical Guide to Opening and Carding – W. Klein, Textile Institute Manual of Textile Technology, 1987
3. Spun Yarn Technology – Eric Oxtoby, Butterworth Publication, 1987, Manchester
4. Elements of Cotton Spinning, Carding & Draw Frame – Dr. A.R. Khare, Sai Publication, 1999, Mumbai

9 Features of Modern High-Speed Draw Frame

9.1 FIBRE CONTROL[1,2]

The intensive research on draw frame sliver quality coupled with the demand for higher production has resulted in spectacular improvements in machine design. The recent developments have tackled the problem of fibre control, and fly and fluff shedding during high-speed drafting so as to achieve the higher speeds without much affecting the quality of the sliver.

During drafting, the fibres pass through the roller nip. But somewhere before or after the roller nip, the fibres are made to pass over (touching) the guiding surface of rotating rollers. This greatly improves control over the fibre movement during drafting. Thus, even when the nip-to-nip distance between the drafting roller pairs is set according to the staple length, the effective distance is reduced. The fibre guidance is thus enhanced. In a modified 3/3 drafting system, the second top roller is tilted in a forward direction, whereas the back top roller is pushed back (Figure 9.1a). There is an apparent increase in the nip-to-nip distance, MN, and it takes care of longer fibres. In Figure 9.1b, however, the back top roller is tilted in the forward direction. In this case, the improved contact of the material over the bottom surface of the second top roller is quite obvious. The material, after having passed through nipping point A, continues to be in contact with the second top roller until point C. Here, the effective controlling distance is reduced from BA to BC. The roller surface between points A and C continues to touch the fibres. This helps in exercising control over the premature forward movement of the shorter fibres which are under the influence of neighbouring accelerated fibres. Allowing the fibres to ride over a roller surface beyond the nipping point seems to have been one of the important features of most of the modern drafting systems. In fact, it may be stated here that keeping the nipping line of top and bottom rollers in vertical direction in any (i.e., conventional) drafting system has a major drawback. In that, when the fibres leave the nipping line, immediately they are under the influence of other neighbouring fibres which are positively pulled by the faster-moving pair of rollers ahead. This, as mentioned earlier, creates a floating fibre condition and leads to drafting wave. In short, in any modern drafting system, many drafting pairs are offset so that their nipping line is not in vertical direction.

In 2/3 drafting (Figure 9.1c), a common top roller, resting over the second and third bottom roller, brings about a similar effect in closing the effective distance in the main drafting zone.

It may be noted here that with the same top roller running over the second and back bottom rollers, there is no draft between them. Therefore, the drafting takes place between the front roller and the second roller (single zone drafting).

DOI: 10.1201/9780429486562-9

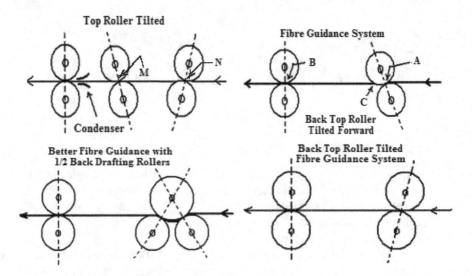

FIGURE 9.1 Drafting systems with improved fibreguidance[3,4]: As against the nip-to-nip fibre gripping in old conventional drafting systems, in all modern drafting systems, the fibre guidance system is improved by extending the roller contact with the material beyond the actual nip. This helps in better control over the shorter fibres. (a) Effective distance widened, (b) effective distance changed, (c) 2/3 drafting system and (d) modified 2/2 drafting.

The normal 2/2 drafting can be modified in the same manner (Figure 9.1d) for longer staple by deviating the back top roller in the backward direction. In both these cases, there is a distinct advantage of the material being guided over the rotating surfaces of either top or bottom rollers. It comprises three lines of top and bottom rollers arranged in a typical way so as to give two distinct drafting zones—break draft in the back zone and main draft in the front zone. Along with the pressure bar (Figure 9.2), the system is designed to run at a very high speed and process a wide range of staple lengths.

The system employs a unique method of fibre control by employing a smaller diameter round bar held at a fixed distance from the surface of second roller pair.

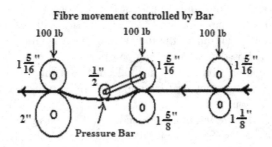

FIGURE 9.2 Fibre controlled by bar[4]: The introduction of a pressure bar has been revolutionary. A very small, round bar that is loosely centred at the middle top second roller is designed to rest on the material being drafted. It is able to restrict unwanted and premature movement of short fibres passing under it. This greatly improves the fibre control.

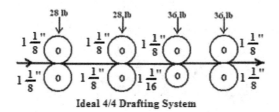

Ideal 4/4 Drafting System

FIGURE 9.3 Ideal draw frame drafting[1,5]: Though the system appears to be like conventional systems, the gripping is improved by using fluted metallic top and bottom rollers.

When adjusting the main zone setting, it is only necessary to alter the relative position of the pressure bar with respect to the second top roller. For short staple, the bar is set forward and the second top roller is also tilted in a forward direction.

For longer staple, the bar is tilted backwards and so is the second top roller. The front top roller in this case is also tilted, but in forward direction.

9.1.1 SPECIALITIES OF SOME OTHER DRAFTING SYSTEMS

1. Ideal draw frame with special metallic top and bottom rollers with flutes on both (Figure 9.3).
2. Saco Lowell Versamatic draw frame[13]with 3 over 4 drafting system can process staple length from 1 in to 2 in. The middle top roller does not touch the second bottom roller. There is a very small gap between the two and this allows *slip drafting* (Figure 9.4).
3. In Zinser draw frame[1,6], there are two top rollers on the front bottom roller. A modified version of 4/5 drafting (Figure 9.5) offers the tubular shaped web collector. For setting the bottom rollers, the second and fourth top rollers are made movable. The draft is split into two zones, the back zone between the fourth and fifth bottom rollers and the front zone between the second and third bottom rollers.
4. In 3/4 RP-600 drafting[1,6] (Figure 9.6), the functioning of each of the machine deliveries is fully independent. Therefore, the setting of drafting rollers on one delivery does not affect the one on adjacent delivery. Also, there are minimum incidences of bottom rollers running out ($< 10\ \mu$). The draft applied is

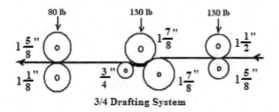

3/4 Drafting System

FIGURE 9.4 Principles of Saco Lowell drafting[4]: The extended fibre guidance is provided by using a common top roller on the second and third bottom rollers.

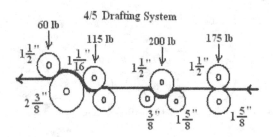

FIGURE 9.5 Principles of Zinser drafting[4]: The fibre guidance is far better in this system. In both front and second zones the typical arrangement of top and bottom rollers provides improved fibre control.

based on the ideal curve of irregularity and this contributes to high evenness. For a normal input sliver, the quality of the sliver produced is in the range of 2.2–2.5 U%.

5. Another version of the 3 over 4 drafting (Marzoli Principle) system is shown in Figure 9.7. It is often said that these versions were developed from the conventional 4/4 drafting system. The main difference, however, is that this, as well as all modern drafting systems basically, are two zone drafting systems—the break drafting zone (1.2 to 1.8) and the main draft (5 to 8).

In almost all modern drawing frames, individual motors are used for driving the rollers. The advantage in this case is that the speeds and therefore the drafts can be steplessly adjusted.

6. Polar drafting[1,6]: It is used in the Lakshmi Rieter (LR) draw frame (Figure 9.8a) has a usual back pair of rollers followed by two sets of 1/2 drafting system. The second and third bottom rollers are smaller in diameter to enable closer setting.

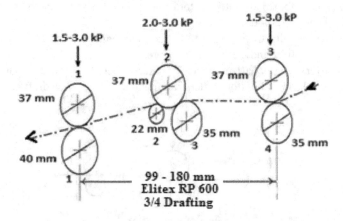

FIGURE 9.6A 3/4 drafting system[4,10]: The fibre guidance is improved in this system because of a common top roller resting on the second and third bottom rollers and overhanging the back top roller in the forward direction.

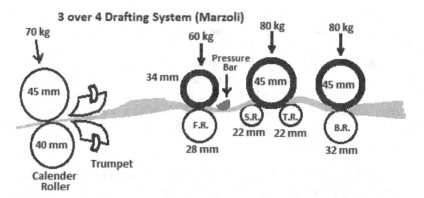

FIGURE 9.7A   3 over 4 drafting—another version[3,4]: The difference between this and earlier systems is that the two bottom rollers under second top roller are evenly balanced. So there is no overhang for the back top roller.

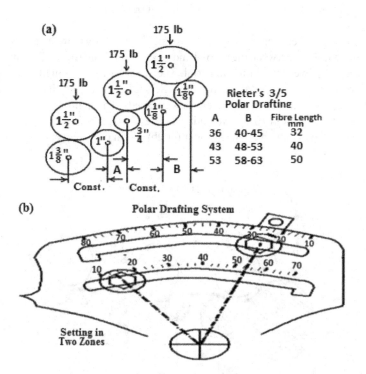

FIGURE 9.8   (a) Polar drafting in LR draw frame[1,6]: A unique design of drafting system. The drafting rollers are arranged in an arc and help to improve fibre guidance. The engraved scales enable direct setting of rollers and (b) roller settings in LRdraw frame: The scales are engraved on curved strips for both back and front zones. With this, it is possible to quickly carry out settings without using the flat roller gauges.

The top rollers' weighting is pneumatic and can be varied infinitely. The air pressure of 0.75 kg/cm^2 is normally used. The rollers are arranged in an arc of a circle, enabling their movement by resetting levers.

The scales are provided (Figure 9.8b) so that the rollers can be set without the use of roller gauges. This arrangement saves a lot of time. Also, it is not necessary to have experts feeling the gauge while passing it between the rollers.

Further, the arrangement also removes any possible anxiety of non-alignment of the rollers, which normally creeps-in when the human element is involved while carrying-out the setting.

7. Principles of Whitin's Acu-Draft draw frame[1,6]: It is again a two-zone drafting system (Figure 9.9) with a 4/5 roller arrangement. After the back pair of rollers there is ½ drafting arrangement in the middle. The setting between the third and the fourth roller is fixed. The use of small-diameter second and third rollers permits closer roller settings for short staple cottons. Though the front and the second bottom rollers run at the same surface speed, the former is made of larger diameter. The settings are made between the second and the third (main zone) and the fourth and fifth (back zone) bottom rollers. The actual front roller speed varies from 180–250 m/min (approx. 600–800 ft/min).

The closest setting between A to B is 1 7/16 in and B to C is 13/16 in. It can also be seen that the main draft is between the second and third rollers. The front roller, in this case, merely acts to deliver the drafted material. The advantage of making the diameter of the front roller largest among all the bottom rollers is again obvious. It helps in increasing the production rate at comparatively slower speeds. The main draft being between second and third bottom roller, weighting on the second and third roller are much higher (200 lb) than that on the front roller (60 lb) which only delivers the material.

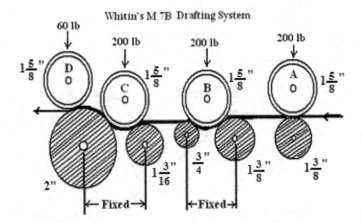

Whitin's M 7B Drafting System

FIGURE 9.9   Principles of 4 over 5 Whitin drafting[5,7]: The main draft is arranged between the second and third bottom rollers whereas the break draft is kept between the fourth and the fifth bottom rollers. The typical placement of front, second and third top rollers helps in improving fibre guidance.

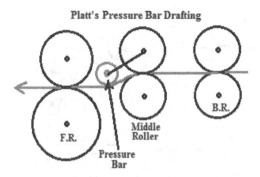

FIGURE 9.10   Principles of Platt's pressure bar system[5,7]: Among all the drafting zones, the fibre acceleration is at the maximum in the front zone. Providing the pressure bar in this zone restricts the premature movement of shorter fibres.

8. Principles of Platt's pressure bar system[2,6]: They have been the pioneers in introducing the pressure bar system in the main drafting zone, where the fibre acceleration is the highest (Figure 9.10).

Introducing a bar with a certain pressure exercised on the fibres is claimed to put a control over unwarranted movement of short floating fibres. This is because, in this zone, the fibre acceleration is the maximum. With the pressure bar in the main drafting zone, the actual settings in the front zone can be done by giving 5–6 mm allowance, whereas the allowance of 9–10 mm in the back zone helps in reducing the drafting force in the break zone.

9.1.2   RIETERS RSB 951[3]

Basically, this is a 3/3 drafting system (Figure 9.11) distinguished by a simple construction. It is easily adjustable, simple to clean and suitable for all fibre types and sliver weights. A pressure bar is provided in the front (main) drafting zone

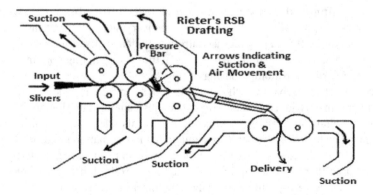

FIGURE 9.11   Rieter's RSB 951 drafting[3,5]: Apart from a pressure bar in the front zone, the unique aspect of this system is the provision of suction to remove floating fibres and dust. It also incorporates an auto-levelling mechanism.

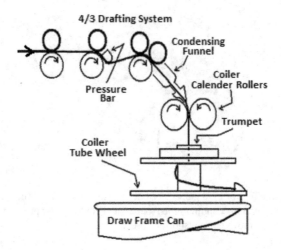

FIGURE 9.12 Trutzschler's TD-7 draw frame[4,8]: In addition to a pressure bar in the front zone, the drafting system provides a higher number of top rollers. Thus, an additional front top roller towards the delivery side guides the sliver very close to condensing funnel.

and offers excellent fibre control. The suction around the drafting system has been distinctly improved.

Thus, an entire area above and below, as well as that around the draw-off rolls, is continuously under the suction. The fibre-to-fibre friction caused during drafting separates the trash and dust particles and these are removed by the suction. It leads to very effective reduction of trash and dust, and thus improves the sliver cleanliness. This seems to be an important aspect in further processing of the material, especially on rotor spinning. Apart from the reduction of the trash and the dust, the suction also serves to keep the draw fame working area clean.

9.1.3 Trutzschler's TD-7 Drawing[8]

This is 4/3 drafting with a pressure bar. The front roller in a tilted position (Figure 9.12) ensures more careful sliver deflection at the delivery side. The adjustable pressure bar in the main draft area provides a controlled guidance of shorter fibres.

The top rollers are guided in the housings of the corresponding bottom rollers and ensure their perfect positioning during setting.

There are cleaning bars which can be used in six positions. The top rollers have special permanently lubricated bearings. They give smooth running and are able to control the temperature rise very effectively. This also helps in preventing lapping formation. The top roller supports have integrated lap formation monitoring, and the smallest possible layer wrapped around the rollers can be quickly detected.

Every time there is a lap-up or even after every roller buffing, this monitoring allows automatic zero-point setting when the machine is restarted. The top roller and the drafting cylinder (bottom roller) form a unit which can be automatically adjusted. The top rollers of the drafting system, as such, are pneumatically loaded.

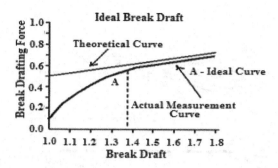

FIGURE 9.13 Optimizing break draft[4,8]: The system is provided with automatic measurement of drafting force for a given mass, type of fibre processed and the force of friction among the fibres, and between the metal and the fibres. The ideal value of break draft is thus indicated.

The suction system is adapted to the drafting system geometry and ensures excellent dust removal from the sliver passing through. The suction hoods are equipped with integrated strippers which are made to act on the bottom rollers and keep them clean all the time.

There is a new groove and sensing roll unit which does auto-levelling, and the sliver evenness is within 0.4 CV%. This results in great reduction in yarn count variation to less than 1%. The servo creel is provided with an individual drive. This helps in eliminating the mechanical connections from the draw frame and relieves high mass moment of inertia. A unique feature of this draw frame is auto-draft optimization.

Here, the break draft is automatically optimized (Figure 9.13). It is known that the break draft is a decisive factor for yarn quality and has significant influence on evenness of the yarn, its strength and imperfections. It also influences the ring frame performance.

The draw frame moves along the entire range of break draft, measures the drafting force in less than a minute and then selects the ideal break draft. The system takes into consideration factors such as - fibre mass, its characteristics, fibre-to-fibre and fibre-to-metal friction, machine settings and even ambient conditions. When installed on one machine, it is possible to transfer this information to the other draw frames.

9.1.3.1 Change in the Main Draft

Opti Set is a standard feature which automatically determines the optimum value of main draft. While doing this, again it takes into consideration parameters such as machine settings, characteristics of the material processed and ambient conditions.

The sensor scans the feed sliver (Figure 9.14). There is a distance of 1m from this point to the main drafting zone where the actual levelling action takes place. A suitable time lag is provided so that when the material arrives in the main drafting zone, the necessary corrections in the main draft are brought about. Initially, when the tenter feeds the sliver, some standard value is chosen for the bulk of the material processed and successive deviating values are checked.

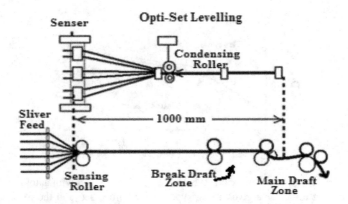

FIGURE 9.14 Sliver levelling[4,8]: There is always an optimum value for the main draft employed in a drafting system. This depends on fibre characteristics, ambient conditions and how close the rollers can be to minimize drafting wave. The time lag between the actual sensing point and the place of corrective action are very important in this case.

Parallel to this process, the CVs of input and output slivers are measured and compared. Based on this, the optimum main draft value is arrived at and displayed. On the touch screen, the tenter can confirm this value by simply touching the screen on the monitor.

9.1.3.2 Servo Draft in TD-8

This provides a high-level of dynamic compensation for deviations from the targeted sliver weight. Control over the short-term evenness is not possible with conventional drafting systems. But in *servo-draft* the special drives used (Figure 9.15a and b) are able to determine the precise values of the drafts in the respective zones.

The measurements involve friction-free and fibre-friendly devices and require high adjustable pressure at the point. Thus, the deviations in the material thickness (fibre–mass deviations) are very accurately measured. Both the sensing and conveying of the signals are highly accurate. These signals are translated and relayed to servo motors to bring about the corresponding changes in the draft. The disc leveller which senses the variation is again a typical grooved roll and sensing roll type, and both are provided with permanently lubricated bearings. They can be replaced when the lots are changed. Based on experience, the values are integrated into software and therefore become very useful for any future reference. The mechanism is even made to function during can changes. All the operations are digitally controlled. Thus, the servo motors are digitally operated. They drive the drafting elements with a toothed belt and do not involve gears or change wheels.

Further, the direct drive and elimination of mechanical gears have a beneficial impact on power consumption. Depending on the application, power consumed ranges between 0.02–0.03 kW per kg of sliver processed. Another interesting feature is that the ideal settings can be stored. They can be later recalled at any time when processing a similar material. The permanently lubricated bearings considerably reduce maintenance and the motors are almost maintenance-free. The suction filter box has larger capacity and provisions are made to prevent any fibre

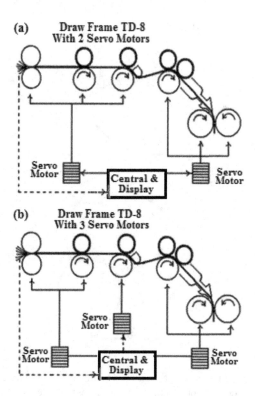

FIGURE 9.15 Servo motors[4,8]: In conventional drafting, it is not possible to precisely control the speeds of the drafting rollers. In modern machines, after the detection and sensing of material mass, conveying it to servomotors brings accurate changes in the speeds of the rollers and hence in the drafts. (a) Two servo motors and (b) three servo motors.

build-up. The spectrogram monitoring ensures that when the machine falls below predetermined quality standards, it is switched-off and spectrogram analysis helps in shortening the trouble-shooting process. This is indicated in the form of an "error" and the possible failing source is marked on the gearing diagram. The working speed for models T7 and T8 is around 1000 m/min and material up to 60 mm can be processed. The draft ranges from 4 to 10 and the suction ranges from 600 to 800 m³/h at -350 to -380 Pa for filters.

9.1.4 INNOVATIVE FIBRE GUIDANCE

It is a sort of an unwritten rule that, in any modern drafting system, there are an uneven number of top and bottom drafting rollers. The placement of these rollers ensures that the material movement around the rollers provides adequate fibre guidance to improve the sliver regularity. In the conventional drafting system, once the slivers leave the nipping line of any pair of top and bottom rollers placed in a vertical line, there is nothing to provide for sliver guidance. By making the top and bottom rollers uneven in their diametric measurements, the material is guided over the roller surfaces and thus premature movement of short fibres is restricted. The earlier

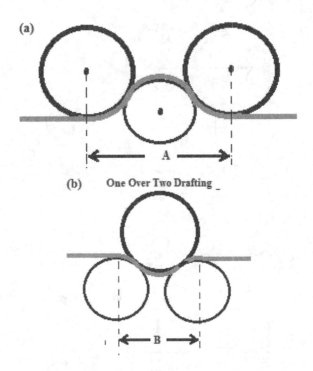

FIGURE 9.16   Fibre guidance comparison[3,4,8]: The distance over which the fibres are supported during drafting depends on whether the number of top rollers or the number of bottom rollers is greater. Some modern drafting systems have preferred and incorporated 2/1 in place of a 1/2 system. (a) Fibre guidance with two over one drafting and (b) fibre guidance with 1 over 2 drafting.

modifications in the drafting system (Figure 9.16b) provided lesser top rollers and comparatively more bottom rollers (e.g., 2/3, 3/4, 3/5 or 4/5)

The innovation came later when the number was reversed (Figure 9.16a) in few modern drafting systems. Thus, the top rollers were more in number than the bottom rollers (i.e., 4/3 or 5/4). Figures (9.17 and 9.18) will give fair idea of the improved fibre guidance with the latter type of versions.

Depending upon the placement and the roller diameters, there is a distinct advantage of using an odd and even number combination of top and bottom rollers. This is revealed by the distance A in a system where the material is made to ride over the bottom roller and under the two top rollers. As compared to this, the distance B in another system providing the fibre guidance is comparatively less. An improved area of contact in A obviously provides better fibre guidance. Based on the system shown in Figure 9.16a, two more drafting systems are described in Figures 9.17 and 9.18.

The additional front top roller 'm', (Figure 9.17), while providing improved fibre guidance, is loaded a little less. This is because the roller does not take part in the drafting and merely delivers the material. However, it is important to note that the fibre guidance between points n and m is much improved. This helps giving in better control over the short fibre movement and thus helps in improving the sliver regularity.

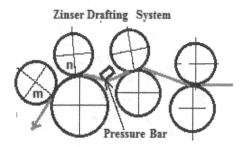

FIGURE 9.17A   4 over 3 drafting system[3,4]: It installed on Zinser draw frame and provides excellent fibre guidance. Apart from roller placement, pressure bar provided in the high draft front zone effectively controls the short fibre movement.

In five over four drafting system (Figure 9.18), all the top five rollers of 39 mm diameter (earlier Rieter system) are pneumatically loaded. The break draft zone is between the third and fourth bottom rollers, whereas between first and second bottom roller, there is a main draft zone. The radial shifting of second and fourth top rollers allows the nip setting to suit the staple length processed.

Very effective and positive fibre guidance is provided by the pressure bar between first and second bottom rollers. The bigger diameter of second bottom rollers (90 mm) provides a larger surface area for better fibre guidance. Another variation with the 5/3 drafting system (Figure 9.19) is basically a two-zone drafting system in which there are two sets of front (A–1 and 2) and second (B–3 and 4) rollers.

In the front zone, there is a 2/1 system wherein there are two top rollers resting on a common bottom front roller. The front top roller is bigger than the remaining top rollers to improve upon the fibre guidance after the sliver web emerges out of the drafting system. It also helps in controlling lapping tendency.

The second roller system, in a similar manner, has two top rollers and a bottom roller. The setting between the front and the second top roller (1 and 2), and that between the third and fourth top roller (3 and 4) is obviously fixed.

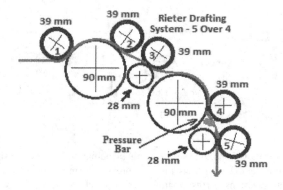

FIGURE 9.18A   5 over 4 drafting system[3,4]: The system exhibits outstanding fibre guidance. Even at high speed, control over the short fibre movement is remarkable.

5/3 Drafting System

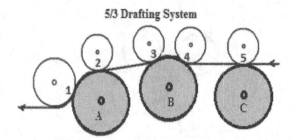

FIGURE 9.19A   5/3 drafting system[3,4]: Yet another drafting system, using more number of top rollers. The two top rollers each on front and second bottom roller provide a much better fibre guidance in front zone where high draft is given.

The main draft is between the two pairs—second top roller and front bottom roller (2 – A) and the third top roller and second bottom roller (3 – B). Similarly, the break draft is between fourth top roller and second bottom roller (4 – B) and fifth top roller and third bottom roller (5 – C). The ingenuity of this system is that the break draft and the main draft zones are thus totally separated. With this arrangement, there is no tendency for the roller to slip. With better fibre guidance in both main and break zones, it is possible to give 3–4 mm allowance in the front zone and 10–15 mm allowance in the back zone and still maintain better sliver evenness.

9.2   MODERN ROLLER WEIGHTING

Unlike conventional draw frames, where the top rollers were weighted by hanging weights, modern methods of roller weighting systems employ top arm spring weighting, pneumatic weighting or magnetic weighting. In spring weighting (Figure 9.20 (a and b), coiled springs are enclosed inside a cylinder and the free end carries a plunger which, in turn, is connected to a bush resting on the top of needle bearings. As mentioned earlier, these needle bearings are carried by the roller necks. Thus, the spring pressure is very precisely exercised on the roller. An arm carrying the spring cylinders for respective rollers is fulcrumed at the back and is put in operation by a

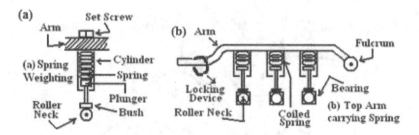

FIGURE 9.20   Spring weighting system[1,2]: This principle of compressed springs is used in weighting the top rollers. The springs in a compressed form are positioned within the top arm at appropriate positions corresponding to the top rollers placed below. When the arm is locked, the springs exert precise pressure on respective top rollers. (a) Varying the spring pressure and (b) top arm with coiled springs.

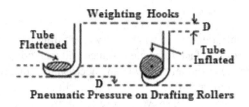

Pneumatic Pressure on Drafting Rollers

FIGURE 9.21 Principle of pneumatic loading[5]: The flattened tubes, when forced in with air under pressure, become rounder and thus exert pressure on the hooks placed around the necks of the top rollers.

locking device. To relieve the pressure at the time of removing the top rollers, this locking arrangement is freed and then the arm is simply raised.

With the arm weighting (Figure 9.20a and b), it is easily possible to clean the accumulated fly and fluff. Further, whenever the rollers are set for processing different staple lengths, it is possible to appropriately shift the coiled spring cylinders which are positioned and fixed by set screws. This arrangement enables the precise pressure of the top arm springs to act accurately on the top of roller bearings.

The pneumatic weighting on the other hand (Figure 9.21), uses air pressure to load the rollers. The principle used here is that the air from the compressor, is let into specially designed tubes which are initially elliptical in shape. When air is forced in, they become more circular, thus pulling down the links.

This results in a pull on the weighting hooks in a downward direction and this exerts the load on top rollers. The arrangement is provided to vary the pressure from $0 - 2.0 \text{ kg/cm}^2$. Usually for cotton, the pressure ranges from 0.75 to 0.85 kg/cm^2. The resultant loading thus realized is equivalent to about 80 kg on the top rollers. When processing synthetic fibres, the pressure is increased to 1.0 kg/cm^2 or sometimes a little more.

In magnetic loading (Figure 9.22), the temporary magnets are formed to put the pressure on the top rollers. Thus, the two magnets, A and B, when locked-in, induce a strong magnetic field to exercise a high magnetic force. This is how the same force is applied for loading the top rollers. The pressure on the top rollers can easily be varied when the magnetic field is changed by varying the current that induces the magnetism.

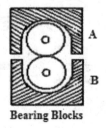

Bearing Blocks

FIGURE 9.22 Principle of magnetic loading[5]: Two electrical magnets (A and B) are used to exert pressure on the bearing blocks. The strength of the magnetic force can be altered by changing the value of the current.

9.3   WEB CONDENSATION

Unlike a conventional draw frame where the web coming out from the front roller is passed over a smooth surface for some distance and then condensed by a trumpet, the high delivery speed in modern draw frame necessitates immediate condensation of web as soon as it emerges from the front roller line. The condensation tube or chute (Figure 9.23) is, therefore, set very close to the front roller nip and the web is condensed very close to front roller nip. In this condensed form, the material passes through the tube and moves directly into the trumpet without disturbing the web formed.

The fibres are thus guided effectively, and the fly and fluff generation is greatly reduced. Some of the latest draw frames have incorporated asymmetric web condensation for reducing the effect of drafting wave. In this case, the condensing tube is kept at the same close position but the web is collected at one side and not centrally.

9.4   SUCTION HOOD

The generation of high levels of fly and fluff is a common problem associated with higher speeds. As against six doublings in the conventional frame, the modern frames can take up 8, or in some cases even 10, doublings. The draft eventually employed in these cases is around 8 or 10. The higher draft working along with higher speed leads to excessive fly and fluff generation. It mostly consists of short fibres which are liberated free. If they are let loose, they would invariably pollute the drawing room atmosphere. Along with the short fibres, there is a small percentage of fine dust resulting from the crushing effect of impurities that are left-over by carding. If they are allowed to be expelled, it would seriously pollute the surrounding atmosphere and this would be quite hazardous to health.

A hood covering the drafting zone and extending right up to the condensing web at the trumpet encloses the whole area. A suction—not very powerful but quite adequate to suck the liberated fly and dust—is provided through this hood and it carries these pollutants away. A fine perforated screen covering the suction fan collects the

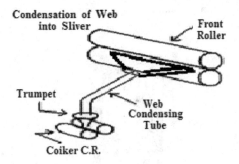

FIGURE 9.23   Condensing tube[4]: On a high speed and high drafting draw frame, it is absolutely essential to partially condense the web coming from the front roller. If this is not done, the web is likely to become unevenly stretched. Also, the tube prevents floating fibre generation.

fly and fluff along with the fine dust. In earlier versions, the tenters were instructed to periodically remove the accumulated matter over the screen. In most modern installations, the whole sucked matter is passed on to filtration unit, and the fly and fluff are automatically sucked away without involving any human element.

Another purpose the hood serves is that it works as safety door. Arrangements are made so that the micro-switch provided allows the machine to run only when the hood is fully closed. When it is lifted, therefore, for any inspection or minding the sliver break at the trumpet, the machine automatically stops. With the hood lifted in this situation the machine can be started (for letting the sliver through the trumpet) only by inching button. After putting down the hood, the machine can be started with the normal speed.

9.5 STOP MOTIONS AND INDICATOR LAMPS

A modern draw frame is equipped with various types of stop motions, viz. sliver break stop motion, full-length stop motion and stop motion when there is roller lapping or sliver choking at trumpet. All of them are electrically operated. When the machine stops for any of the above reasons, a lamp of a certain colour glows. The different colours indicate different reasons for the machine stoppage. This enables the tenter to know the reason for such stoppages even from a distance. Thus, a yellow or orange lamp may be used for sliver breaks, whereas a red lamp may indicate roller lapping. A green lamp is usually used for a full length coiled into the draw frame can. If can changing is manual, the tenter may replace the full can with an empty one and reset the counter before restarting the machine. With automatic can changing, all this is done automatically. In such cases, the worker has to merely arrange for the empty cans on circular can changing plates mounted on a turntable. In all modern machines, the full length stop motions work on coiling a pre-determined length into the can. The conventional machines working without measuring counter were often overstuffed with sliver. In modern draw frames, when the full-length stop motion pauses for a can change, it is for only a short interval. When an empty can automatically replaces the full can, the machine restarts by itself. The tenter is then expected to remove the pushed-out full can. On latest modern draw frames, the machines are made to work on slow speed during auto-can changing operation.

When the machine is running without any interruption, a white light continuously glows. Although each worker looks after only two sets of machines in modern sequence, the work of the tenter is never a simple one. He has to mend the break when sliver breaks at the back or at the front, clear the roller laps and even clean the rollers when lapping becomes frequent. He has to clear the fly and fluff gathered over the suction screen (in earlier versions of high-speed draw frames) and then replace the delivery cans. He is expected to clean the machine from time to time and keep it tidy. Though a major part of this work is reduced in modern draw frames, the provision of indicator lamps helps him considerably in locating faults. He is able to quickly deal with them. With such stop motions, when he is attending his assigned machines, he can work freely and hence more efficiently. This results in higher productivity with better quality.

The stop motions are also used as a safety measure. The head stock doors of all modern machines are provided with stop motions. When the doors are opened for any reason, a micro switch immediately brings the machine to stop. Only when the doors are properly locked can the machine start. Thus, when the repairs are being carried out, any accidental starting of the machine is totally avoided.

An inching motion provided on the machine enables running of the machine for very short duration. This becomes necessary when the worker takes the web delivered by the drafting rollers to pass it through the coiler. For doing this, he has to open the suction hood above the drafting system. He then uses the inching button and brings the sliver ahead. Once the end of the condensed sliver is picked by the coiler calendar roller, the worker joins the two ends of the slivers in the can. The hood is replaced. Only then can he push the normal starting button to work the machine at full speed. The *on, off* and *inch* buttons are provided on the front as well as at the back of the machine so that the tenter can operate it from both sides. This is very convenient, especially when he carries piecing of broken sliver at the back creel.

9.6   EASE IN ROLLER SETTING

On modern draw frames, the roller setting operations are made simple and yet precise. It is not necessary to put the flat gauges in between the rollers to set them correctly. The bearing blocks are made in such a way (Figure 9.24a) that when a certain distance is set between them, it automatically brings about the required gauge between the rollers. This reduces considerable time in ensuring whether the rollers are perfectly parallel or not. Usually the manufacturers recommend distances in between the blocks depending on the staple length processed and simplify the matter. The common choice is a *4–8* (4 mm and 8 mm) setting which may be used for the first drawing passage, whereas *3–6* may be chosen for the second passage. The post

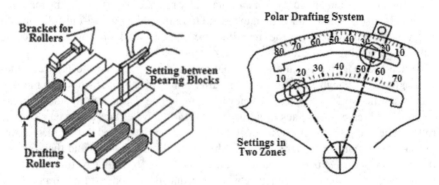

FIGURE 9.24   Roller settings on modern draw frame[3,7]: The tedious procedure of inserting the flat gauges between the rollers, the provision is made in modern drawing frames to quickly and yet very accurately set the rollers. (a) Whitin draw frame[7] and (b) Rieter draw frame.[3]

comb drawing frame, owing to better parallelization and removal of short fibres in the earlier process, may have a still finer setting.

On the Rieter draw frame equipped with *polar drafting* system (Figure 9.24b), an ingenious method of roller setting was used. The third and fourth rollers are held in one bracket, whereas the fifth is held in yet another bracket. These are carried by the levers on either side and are capable of being moved independently. The engraved scales are provided with markings in millimetres so that if the two brackets are set on marks *30–33* it means that the front zone setting is 30 mm whereas the back zone setting is 33 mm. These settings can be done quickly and accurately without the need for insertion of any flat gauges.

Yet another advantage with such a provision is that the respective rollers are perfectly aligned when the setting is carried out. So, there is no roller flexing.

9.7 BIGGER CANS AND SPECIAL COILERS

With high production rates, it becomes necessary to use bigger packages (cans) both at the creel and at the delivery. The bigger cans at the creel minimize the can replacements at the back and thus improve operator efficiency, whereas those in the front accommodating more material improve machine efficiency. A little earlier, the can size of 400 mm × 900 mm (15 in × 35 in) was very popular and extensively used. However, much larger cans of 1000 mm × 1500 mm (40 in × 59 in) are available on some of the recent modern draw frames. The bigger base for the cans also makes them more stable. The latest design of can provides casters at the bottom so material transportation becomes easy. When using can changers, especially with 1000 mm diameter, they can be sunk into the floor so that it is possible to remove the cans at ground level after a can change. The use of the bigger cans necessitates a coiler of bigger size.

In addition, they are equipped with a *twin* or *bi-coiling* sliver system. In a twin system, the total slivers fed to the drafting system are divided into two groups. The two webs thus emerging are consolidated and combined by a special trumpet but are coiled into a can as one sliver. The cans are rotated in alternate directions after each rotation to avoid slivers getting twisted. The important advantage in this case is that it reduces the amount of draft in the drafting system.

In bi-coiling, the two emerging webs are condensed by two separate trumpets but are condensed in one can only. The coils are distinct from each other and the can rotates only a half revolution in either direction. With both systems, the number of cans at the creel of the next machine is reduced and this reduces the creel space considerably. The tenter is thus able to work more freely and efficiently.

9.8 ROLLER DIAMETER

As mentioned earlier, the roller weighting on a modern draw frame is considerably higher, thus necessitating strong and sturdy rollers. The bottom rollers, especially, must be capable of sustaining higher pressure. They are therefore not only made bigger in diameter, but also made of steel. This reduces any possibility of their bending or buckling under the top roller pressure and thus becoming eccentric.

The modern draw frames invariably have the front bottom rollers comparatively bigger in size. There are two reasons for this. Firstly, the pressure on the front line rollers is much higher. Secondly, the bigger diameter of the roller gives more production in terms of the delivery rate for the same speed.

9.9 AUTOMATIC REGULARITY CONTROL[3,4]

The draft employed in the drafting system decides the hank of the sliver produced. With eight or ten slivers fed, the corresponding draft of eight to ten respectively should make the hank of the delivered sliver almost the same as that fed. However, apart from other machinery factors like roller weighting, roller vibrations or eccentricity, the fluctuations in the input hank sliver directly affects the variation in the delivered material. In short, the thinning-out process carried out in a drafting system will be proportional to the overall weight being fed. Therefore, a finer overall hank of the slivers fed at any instant is expected to result in a finer hank delivered and vice-versa. This leads to constant fluctuations in the delivered hank.

The concept of automatic regularity control involves variable draft corresponding to the fluctuations in the input material. For example, when a heavier material is fed, the draft level is proportionally increased so as to maintain the weight of the material delivered.

The whole operation requires a constant sensing of the bulk so that, once sensed, the signals could be relayed to the speed variation mechanism quickly and automatically. The speed of the concerned roller is then altered so as to change the draft appropriately. In some systems, the speed of the front roller is changed, whereas in some others the back roller speed is altered. However, the corrections required in changing the draft in these two systems will have to be exactly the opposite. In the electronic type, the servo mechanism operates in two ways. The principles are based on open loop and closed loop circuits.

9.9.1 OPEN LOOP

In open loop, the corrections are made by changing the delivery rate. It fairly controls the short-term variations. The flow of the material is shown in Figure 9.25.

The measurements of input material (feed) are made, and accordingly, the signals are sent to the *electronic control unit*. These signals are compared by the *control unit*

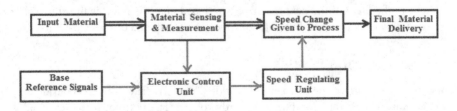

FIGURE 9.25 Open loop principle[2,5]: One of the forms of automatically controlling the sliver regularity. The sensing, the relaying and bringing about the change have been made possible with the introduction of servomotors.

with *base reference signals* which represent the mean hank of the sliver. The control unit, in turn, either increases or decreases the output of the speed regulating unit (regulator). The signals are finally relayed to a variable speed-change device which ultimately provides the variable speed to front or back roller. This changes the draft in the system.

The change in the draft is in proportion to the deviation of sensed material from mean hank. As the material is sensed before it enters the draft zone, the real change in the draft is withheld till the material actually arrives in the draft zone. As mentioned above, depending on the system design, either the speed of back roller or front roller is changed to set the hank as close as possible to the mean hank.

It can be easily seen that, if the direction of the arrows is followed, taking any arrow shown in Figure 9.25, it ultimately leads to the sliver delivery where the arrow finally leads to *open*. The main drawback with this kind of correction, however, is that it is not possible to ascertain how much the change in the draft has brought the correction closer to the nominal hank of the sliver. This is because sensing of the outgoing sliver is not done. In spite of this, the open loop system of auto-levelling is comparatively much simple and easy in design.

9.9.2 Closed Loop

In closed loop, the sensing of the variations in the sliver is done at the delivery. Depending on the variations in the delivered material, the signals are sent to the electronic control unit. These signals are compared with base reference signals. Accordingly, the change in the speed (if needed) is relayed to the speed regulating unit. Ultimately the appropriate change is made in the rate of input material. This makes the appropriate changes in the hank of the material in the process. As can be seen from the diagram (Figure 9.26), excluding the final delivery, the directions of all other arrows lead to a never-ending loop.

The measurements of the material in the process are made at the feed side in the open loop system, whereas they are made at the delivery side in a closed loop. It is often criticized that in a closed loop the sensing is done on the material already delivered, whereas the corrections are made on a totally different input material during drafting. But it must also be remembered that this corrected material coming from the input side is again going to be sensed and compared with the reference signals when it is finally delivered.

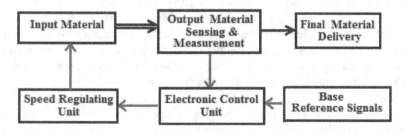

FIGURE 9.26   Closed loop principle[2,5]: Another form of controlling sliver regularity. The output material is sensed to bring about the change in the draft in the earlier zone.

It is known that open loop is responsible for regulating short-term variations in the sliver. As against this, medium- and long-term variations are regulated in closed loop.

Another point of interest is that in open loop a much thicker mass of the feed material, moving at a comparatively quite slow speed, is continuously sensed and measured. This gives sufficient time for the sensing device in an open loop to carry on its function of measurements. Thus, it enables the sensing device to achieve more accurate and precise sensing.

Whereas in close loop, the sensing and measuring device is set on the delivery side and the material is comparatively quite thin. Further, the thickness measurements must be made on a material moving at a much faster rate. Thus, the time for sensing and measurement is quite short. This situation is very likely to lead to a slightly less accurate measurement.

9.9.3 COMBINED LOOP

Open loop and closed loop systems can be combined to give their joint benefits. Thus, a system may sense and measure the hank of the input sliver (as in open loop) and also finally check the hank of the delivered sliver. It is thus possible to maintain the sliver delivered precisely very close to the nominal sliver hank. It is also possible to measure the sliver hank (thickness) as it is being drafted.

9.9.4 IMPORTANT CONSIDERATIONS

Basically, the draw frame holds a key position in the material flow through various spinning stages. The variations in the material fed to the draw frame are of short- and medium-term nature. As the material from draw frame subsequently is drafted further, these variations reflect as medium- and long-term variations in the final yarn. Also, at all the stages of spinning, the number of machines involved in carrying out their respective jobs is quite large as compared to the number of corresponding draw frames. Thus, *auto-levelling* can be most beneficially applied at the draw frame stage. However, auto-levelling cannot possibly curb the irregularity arising out of random fibre variation.

It is also interesting to note that when the back roller speed is changed for auto-levelling variations, there is no change in the production. This is because; the production rate depends upon the surface speed of final delivery roller. Also, as the back roller runs at a comparatively slower speed (as compared to the speed of front roller), there is much less disturbance to the machine functioning. When the front roller speed is, however, changed (higher speed) for correcting variations, it creates jerks in the gearings. Also these jerks are passed-on to the heavy coiler system, and this leads to machine vibration. Also, as the production depends upon the front roller speed, every corrective action in the front roller speed influences the production.

It is customary to employ auto-levelling as late as the draw frame stage. With two passages of draw frames therefore, it is always advisable to put the controlling unit on the finisher draw frame. All single delivery finisher draw frames of Rieter are equipped with an open loop system. Usually, after draw frame there is only one intervening process—fly frame—before the final spinning of yarn. Therefore, the risk of introducing of short- or medium-term variation after draw frame passage and before final yarn spinning is considerably less. Though auto-leveller does not eliminate the variations completely, the overall evenness of the sliver and therefore ultimately that of the final yarn is significantly improved. On all Rieter machines, a *tongue and groove* mechanism is used to detect variation in the input material. When the sensing device is installed on the delivery side, it is usually coupled with the trumpet.

9.9.5 OPEN LOOP CONTROL IN RSB 951

The basic function of any auto-levelling device is to reduce disturbing variations in the mass fed. Rieter's 3/3 drafting system (Figure 9.27) in their RSB 951 single-head model, uses the open loop principle to successfully control the variation in the sliver feed. It is claimed that open loop control allows levelling-out of even the smallest deviations in the mass.

Here, the tongue and groove pair measures the variation and stores it until the measured slivers reach the main drafting field. At this very moment, the draft is changed by a highly dynamic servo drive.

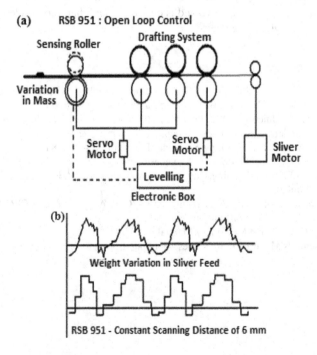

FIGURE 9.27   RSB (Rieter)[3,4]: The open loop principle used in this system offers much better control over the sliver mass variations. The sensed variations are withheld till the material reaches the main drafting zone. Only then the drafting speeds are changed accordingly to correct the variation. (a) Open loop and (b) constant scanning.

These are such drives that immediate change the draft. In this draw frame, the arriving sliver mass is constantly scanned (Figure 9.27) at a distance as close as 6 mm. This makes the scanning independent of the sliver speed and guarantees that the sliver deviations are registered very precisely even at higher delivery speed. In other systems, the sliver is measured at a uniform time interval. It means that the length of the sliver between the two signals becomes large as the feeding speed increases.

For other draw frames, at a delivery speed of 800 m/min, the feeding rate is 130 m/min and the signal distance is around 25 mm. This is almost 4 times larger than the one judged by RSB 951 (6 mm). The draw frame is also provided with a *Sliver Monitor Plus* system, where the sliver thickness variation is measured at the output. It provides all the relevant data and constantly compares the actual hank value with that of nominal. If the actual value exceeds the nominal value by more than a certain limit, the machine is automatically turned off.

In this system, a summary of technical data such as CV%, number and time of machine stops, efficiency over a shift, CVL (1 m, 3 m, 5 m, etc.), is also provided. With one delivery draw frame, it is observed that the efficiency is improved by 10%. A single delivery frame facilitates easier transportation of cans and is more flexibly integrated into the spinning line. The two-delivery draw frame, however, has one distinct advantage— saving space. But when two single-delivery draw frames are placed closer, the extra space involved is only 8.5% and that, too, is adequately compensated for by higher efficiency.

REFERENCES

1. Manual of Cotton Spinning – "Draw Frames, Comber & Speed Frames" – Frank Charnley, Textile Institute, Butterworth Publication, 1964, Manchester
2. Spun Yarn Technology – Eric Oxtoby, Butterworth Publication, 1987, Manchester
3. Rieter's Modern Draw Frames – Pamphlets, brochures & booklets
4. Rieters Manual of Spinning-Spinning Preparation – W. Klein, Textile Institute, Manchester Vol. 3, 1987 (2008), Manchester
5. Elements of Cotton Spinning, Carding & Draw Frame – Dr. A.R. Khare, Sai Publication, 1999, Mumbai
6. Fundamentals of Spun Yarn Technology – Carl A. Lawrence – CRC Press, 2003, London, New York, Washington DC
7. Whitin Draw Frame Machinery Manual
8. Trutzschler's Modern Draw Frames – Pamphlets, brochures & booklets

10 Faults in Draw Frame[1]

10.1 ROLLER GEARING FAULTS

The sliver coming out from the first draw frame passage usually has periodic variations of short-term and long-term wave lengths. In addition, the machinery faults, if any, at draw frame normally contribute their share to short-term variations already present. During the subsequent passage/s, the draft employed changes the variation pattern. For example, a periodic wave of short length changes itself to medium-length variation. It is expected that, in order to obtain a regular draw frame sliver, the mechanical functioning of the machine must be absolutely up to the mark. Some typical faults due to defective mechanical working are discussed below.

As the bottom rollers are all positively driven through gearing, any fault in the gearing affects their speeds. Sometimes the gear wheels mounted on the shafts become loose or eccentric. Either the bore in the wheels or the shaft becomes excessively worn-out. This adds to uneven running of the gears and leads to a periodic variation arising due to corresponding bottom roller. Improper meshing of the gears, broken tooth or teeth in the drafting gears, teeth clogged with fly or fluff, worn out gears owing to inadequate lubrication, strained bottom rollers due to improper alignment—all are common faults which ultimately result in interrupted running of bottom rollers and, depending on their periodicity, introduce a variation of a certain wave length.

10.2 TOP ROLLERS[1,2]

The top rollers are covered with synthetic rubber. However, worn out or bad roller coverings often lead to an uneven grip. This seriously impairs the drafting operation and can lead to roller slip. The *hollowing* or *channeling* is another peculiar defect which is due to wearing of that portion of the roller covering over which the sliver material passes continuously. An efficient traverse motion may obviate or prolong this condition. However, a periodic buffing becomes necessary to smooth out the top roller surface over the cot. This makes the surface, smooth, uniform and even. The hollowing may not be very critical with more bulk passing through the roller nip; however, it may become serious, especially at the front roller, where the bulk is minimum. In such cases, the roller pair fails to exercise a sufficient grip.

The movement of the top rollers in their bushes, improper lubrication at the roller necks where the bearings support them, worn-out needle bearings and improper seating of top rollers over the bottom rollers—all lead to uneven and improper running of the top rollers and thus lead to variation in the drafted strand.

DOI: 10.1201/9780429486562-10

10.3   ECCENTRIC BOTTOM ROLLERS[1]

The bottom rollers, when perfectly straight and properly aligned, ensure a correct length of the material delivered. However, when they are bent, and this usually happens in the middle of the rollers, their running becomes eccentric.

This is a very typical phenomenon and it happens when rollers of longer length are used with their supports placed at much wider distances. The heavy top roller pressure used in this case has a tendency to bend these bottom rollers at their middle portions. This is further aggravated when the rollers are too weak to take-up the higher loading. In fact, this is one of the reasons the bottom rollers in modern draw frame are made larger in diameter.

During their handling at the time of scouring (treatment with alkali-soda) and cleaning, often care of the top rollers is required to be taken. At the time of roller laps, if the machine does not stop quickly, the rollers are pressed excessively. This exercises extra pressure in the central region of the bottom rollers and leads to their permanent deformation—bending. The eccentric running of bottom rollers introduces a periodic loss of the grip between the corresponding pair of rollers and seriously impairs the drafting action. Thus, it introduces a periodic variation in the material.

10.4   ROLLER VIBRATIONS[1]

Machine balancing and the roller vibrations are closely related. At the time of machine erection, it becomes an important criterion to prepare a sound footing at the base where the machine is subsequently bolted. The firm footing at the base helps to considerably reduce the machine vibrations. Even the speed at which the machine is designed to run has to be considered. The speeding-up procedure after the new installation of the machine needs to be carefully followed to ensure smooth running of the moving parts.

Apart from this, another cause for roller vibration is the faulty driving of drafting rollers. When the rollers are mounted during erection, or when they are set, it is often necessary to see that not only are they aligned correctly but also that they run smoothly in their bearings. The erector or installer often disconnects the main gearing and sees this by turning the bottom roller sections by hands. Any stiffness left here seriously impairs their free running. In all modern drafting systems, the setting arrangements of the rollers are improvised so that, even after the change in settings, the rollers still run freely in their bearings. The driving of the front roller from one end and then reaching this drive to the remaining rollers from the other end also gives rise to roller stiffness leading to vibrations.

10.5   BOUNCING OF THE WEIGHTS[1]

With the hanging type of weighting system in the old conventional draw frames, the machine and roller vibrations proved to be detrimental to the normal functioning of the machine. Along with these vibrations, the hanging weights also used to dance up and down. Especially when the weights jumped up, there was a temporary and partial relieving of the roller loading, resulting in the top rollers slipping over their counterparts. The partially relieving of the roller load was also caused when the weighting hooks, during the vibrations, sometimes touched any of the machine parts in the vicinity. This led to the variations in the drafting process and reflected in the quality of the material delivered. In modern machines, the use of top arm spring weighting or pneumatic loading on the top rollers totally removed this lacuna.

10.6   SINGLES AND DOUBLES[2]

Whenever the sliver breaks, the machine must stop immediately. An efficient stop motion can do this provided it is maintained properly. It is often found that the stop motion is either faulty or tampered with by the workers minding the machine. This allows the machine to run even when the sliver breaks, resulting in *singles*. Sometimes, the improper piecing of the broken sliver at the preceding process gives way on the draw frame creel but the stop motion fails to detect the break. This is because the broken sliver holds itself under the break-sensing device. This also results in singles.

Doubles often occur due to inefficiency and carelessness on the part of the tenter minding the machine. After the break, he often tends to avoid a proper piecing and simply guides the fresh sliver on the table along with the other slivers. The extra superimposition of the part of the sliver riding over the others is, in fact, an added quantity resulting in a thick portion. Sometimes the tenter adds an extra length of the sliver from the spoiled material cans, preferably to consume the reusable waste material. This is because it is the duty of the supervisor to periodically check the quantity of soft waste produced by the tenter. Frequent breaks at the creel usually give the tenter this opportunity to reuse this soft waste. Proper training of the workers along with good supervision can reduce occurrence of such events.

There is another reason for the doubles, the source being at the card itself. This occurs at the time of replacing exhausted lap with a new one. In some instances, the workers tend to keep as much as ½ –1 m lap length of overlap of the finishing portion of the lap being exhausted and the starting portion of the new lap. This is not only sheer negligence but it is done by the worker to obtain some extra leisure time. The extra superimposition invariably leads to a much longer portion—as much as 50–100 m of card sliver (draft in the card is around 100)—with almost double the thickness. Even in a chute feeding system at the card, failure to maintain a certain fixed height of cotton also leads to constant fluctuations in the sliver produced at the card. All such thicker slivers invariably act as doubles at draw frame.

10.7   UNEVEN DRAW FRAME SLIVER[1,2]

Basically, the fault can be located around the drafting region. The modern draw frames do provide sophisticated controls over the draft and its distribution, precision in roller setting and roller loading. When the machine is new, all such factors are at their best. However, the machine maintenance and its up-keep are factors which require continuous attention once the machine is installed. Apart from routine oiling and greasing schedules, which may possibly ensure the

smooth running of the machine, the condition of the roller cots, roller eccentricity and periodic checking of roller settings in relation to the change in the mixing are very much required to be watched. The values of the tension draft and
occurrence of *false draft* have to be seriously checked from time to time. This
is because, in carrying-out the routine maintenance practices, the staff involved
may neglect these aspects. Even the trumpet in the coiler must be periodically
checked whenever there is a change in the hank or bulk of material.

A factor often neglected is the spring pressure in the cans. A broken spring or sometimes a card can filled even without springs can lead to increase in tension drafts at the
creel. It is really worthwhile to stress the importance of controlling the tension drafts,
especially between front roller and calender roller where the nipping points are spaced
distance apart, i.e. at a much wider distance as compared to normal roller settings. The
draft has to be certainly the bare minimum in such cases. Equally important is to maintain the stop motions in absolutely good condition. In such cases, the role of the supervisor
is crucial. When in the department, the supervisor must regularly check the effectiveness
of different types of stop motions, and if needed, make them functional.

In conventional set-up, the sliver was taken to the laboratory—and that only
occasionally—to find the Uster value. There are good electronic instruments of
the capacitance type or photo electric type or optic type (Uster, Fielden Walker
or Zweigle) to give accurate and precise evaluation of sliver evenness. They often
aid the technologist to categorize the sliver produced as even, average or uneven.
With them, the quality of the sliver produced can be easily judged. The standard norms are published by the manufacturers of these instruments from time
to time and are of great help. For example, a value of less than 1.5 U% for the
drawn sliver may be considered to be quite good, while a value above 3.0 U%
may be totally unsatisfactory. Even then, collecting the samples, taking them to
the laboratory, testing the sliver and finding the evenness values take some time.
Therefore, the corrective action is always delayed. Further, a technologist in the
mill has to judge several factors before forming his own standards. The standards
quoted by the manufacturers or research organizations often pertain to the ideal
working conditions. The mill has to form their own standards which, according
to the machine conditions and the affordable maintenance practices, are achiev-
able. Against this background, the modern sophistication in online presentation
of the sliver irregularity data seems to have greatly helped the technologists in
quickly finding the variations present in the running sliver. Provisions are also
made on the machines to speedily take corrective action.

As mentioned above, along with the conditions of the machines and the maintenance schedule, the condition of the material being supplied to draw frame, the type
of the product that is finally aimed at and its use are all very important factors that
must be into consideration before forming such standard norms.

In some typical cases, a mill continues to work with an old set of machines which
themselves are sources of variation. If the financial position does not permit complete modernization, either partial modernization has to be sought or the norms must
be made more flexible.

With a short processing, the draw frame has become the center of attention. This is because it is possible to improve both the parallelization and the

uniformity of the sliver material on this machine. Thus, though the draw frame appears to be a comparatively simple machine to understand, its proper upkeep becomes very important and has to be meticulously followed so as to obtain what is expected of it.

REFERENCES

1. Process Control in Spinning, A.R. Garde & T.A. Subramanian, ATIRA Silver Jubilee Monographs, ATIRA Publications, 1974
2. Elements of Cotton Spinning, Carding & Draw Frame – Dr. A.R. Khare, Sai Publication, 1999, Mumbai

11 Card Gearing Calculations

11.1 GENERAL GEARING PLAN

All the calculations are with reference to the gearing diagram for conventional semi-high production (SHP) card and are given in this chapter.

So, both the card and the draw frame calculations are with reference to production, speeds of important parts and drafts. Few of them are related to requirements of machinery to balance them in the production line.

11.1.1 SPEEDS OF VARIOUS PARTS

1. Speed of cylinder $= \dfrac{860 \times 4}{18.3} = 188$ rpm

2. Speed of licker-in $= \dfrac{188 \times 17.5}{5} = 658$ rpm

3. Speed of doffer $= 658 \times \dfrac{5\frac{3}{4} \times 22 \times 36}{12\frac{1}{2} \times 100 \times 216} = 11.1$ rpm

4. Speed of feed roller $= 11.1 \times \dfrac{30 \times 21}{40 \times 120} = 1.46$ rpm

5. Speed of lap roller $= 1.46 \times \dfrac{24}{68} = 0.52$ rpm

6. Speed of calender roller (CR) $= 11.1 \times \dfrac{216}{30} = 79.92$ rpm

7. Speed of tube wheel $= 79.92 \times \dfrac{38 \times 20 \times 26}{18 \times 20 \times 130} = 33.74$ rpm

8. Speed of doffer comb $= 188 \times \dfrac{21\frac{1}{2} \times 11\frac{5}{8}}{5\frac{3}{4} \times 3\frac{5}{8}} = 2254.3$ strokes/min

9. Speed of can plate $= 79.92 \times \dfrac{38 \times 20 \times 20 \times 1}{18 \times 20 \times 40 \times 93} = 0.90$ rpm

10. Speed of flats:

    One revolution of sprocket wheel (Figure 11.2) moves the flats by 15×3.5 cm (owing to its connections with flats).

    Sprocket revolutions per minute will be

    $$\dfrac{188 \times 16.5 \times 1 \times 1}{25 \times 16 \times 40} = 0.19$$

DOI: 10.1201/9780429486562-11

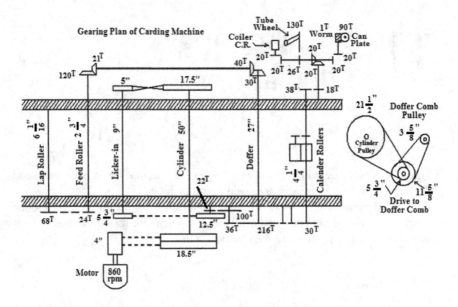

FIGURE 11.1 Gearing plan of carding machine[1]: The specialty of this gearing is the drive imparted to the important organs like cylinder, doffer, licker-in, etc. The drive also includes the two oscillating motions given to the flat comb and doffer comb.

Hence, the flat speed is 1

$$5 \times 3.5 \times 0.19$$
$$= 9.97 \text{ cm/min} = 3.97 \text{ in/min}$$

11. Speed of Coiler CR $= 79.92 \times \dfrac{38 \times 20 \times 20}{18 \times 20 \times 20} = 168.72$ rpm

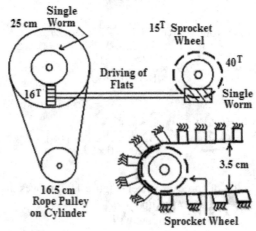

FIGURE 11.2 Drive to flats[1,2]: The flats run very slowly and take part with the cylinder to produce very effective carding action. The required speed reduction for the flats is achieved by using worm and worm wheel mechanisms.

11.1.2 DRAFT CALCULATIONS

1. Draft between feed roller and lap roller $= \dfrac{1.46 \times 2\frac{3}{4}}{0.52 \times 61/16} = 1.27$

2. Draft between licker-in and feed roller $= \dfrac{658 \times 9}{1.46 \times 2\frac{3}{4}} = 1474.96$

3. Draft between cylinder and licker-in $= \dfrac{188 \times 50}{658 \times 9} = 1.58$

4. Draft between doffer and cylinder $= \dfrac{11.1 \times 27}{188 \times 50} = 0.032$

5. Draft between CR and doffer $= \dfrac{79.92 \times 4\frac{1}{8}}{11.1 \times 27} = 1$

6. Draft between coiler CR and CRc $= \dfrac{68.72 \times 2}{79.92 \times 4\frac{1}{8}} = 1.02$

7. Total draft $= \dfrac{168.72 \times 2}{0.52 \times 61/16} = 107.03$

8. Draft constant
$$\begin{aligned} &= \text{Draft} \times \text{Draft change pinion} \\ &= 178.03 \times 21 = 3738.63 \end{aligned}$$

9. Doffer speed constant $= \dfrac{\text{Doffer speed}}{\text{Barrow wheel}} = \dfrac{11.10}{22} = 0.50$

Or

$$= 658 \times \dfrac{5\frac{3}{4} \times 22 \times 36 \times 1}{12\frac{1}{2} \times 100 \times 216 \times 22} = 0.50$$

11.1.3 PRODUCTION CALCULATION

Considering the same gearing diagram, the speed of the coiler calender roller is found to be 173 rpm. The diameter of the coiler calender roller is 2 in. If the hank of the sliver delivered is 0.15 (English hank), the production can be calculated as follows:

$$\text{Production/shift at 90\% efficiency} = \dfrac{168.72 \times 2 \times \pi \times 60 \times 8 \times 90}{12 \times 3 \times 840 \times 0.15 \times 100}$$
$$= 100.97\,\text{lb or } 45.79 \text{ kg}$$

$$\text{Production constant} = \dfrac{\text{Production}}{\text{Barrow wheel}} = \dfrac{45.79}{22} = 2.08$$

$$\text{Production per revolution of doffer per shift} = \dfrac{45.79}{11.10} = 4.12\,\text{kg}$$

Note: This method is uncommon; however, in the regular mill working, where the average doffer speed can be calculated without a tachometer, the approximate expected production can be quickly obtained.

11.1.4 LENGTH OF ONE COIL AND TWIST IN THE COILED SLIVER

$$\text{Length of 1 coil} = \frac{\text{Length delivered by coiler CR per minute}}{\text{No. of revolutions of tube wheel per minute}}$$

$$= \frac{168.72 \times 2 \times \pi}{37.88} = 27.98 \text{ in}$$

$$\text{Twist Inserted in the card sliver per inch} = \frac{\text{Speed of the can plate per minute}}{\text{Delivery of coiler CR per minute}}$$

$$= \frac{0.90}{168.72 \times 2 \times \pi} = \frac{0.90}{1060.23} = 0.00085$$

Note: Though this twist is absolutely negligible, it is important to understand that the card sliver during coiling does get twisted. It may be important in the case of bi-coiling.

11.1.5 DOFFER COMB OSCILLATIONS

When the speed of a doffer comb is higher than what is necessary, this puts a lot of strain on the comb box. The strokes that the doffer comb makes should be just sufficient to strip the doffer. This saves unnecessary wear and tear of the comb box parts due to excessive speed.

If the doffer speed is N rpm and the doffer diameter is D inches, then the doffer surface speed to be stripped per minute will be $\pi \times D \times N$ inches. (11.1)

If the stroke of the comb is Y inches and only f percentage of the stroke is useful, then, per stroke, the comb will peel-off the web of length $(Y \times f)/100$ inches. (11.2)

Note: It can be very well imagined that only a part of the doffer comb would be useful in peeling-off the web.

In addition to downward movement of the comb, it also moves a little distance away from the doffer at the end of its stroke. If this distance is d inches, then every time the comb oscillates, an extra d inches of web will be peeled off.

Therefore, the total length peeled-off = $(Y \times f)/100 + d$ (11.3)

$$\text{The required strokes of the comb box} = \frac{\pi \times D \times N}{(Y \times f)/100 + d}$$

11.1.5.1 Doffer Comb (Worked Examples)

1. A card is running with doffer speed of 14 rpm and the stroke of the comb is $1\frac{3}{8}$ in. Only 40% of the stroke is useful. Further, the comb moves $\frac{1}{8}$ in away from the doffer at the end of its stroke. What should be the comb box speed?

$$\text{Doffer surface speed} = (\pi \times 27 \times 14) = 1187.67 \text{ in/min}$$
$$\text{Web peeled-off per stroke} = (1\frac{3}{8} \times 0.4) = 0.55 \text{ in}$$
$$\text{Extra web peeled-off per stroke} = \frac{1}{8} \text{ in} = 0.125 \text{ in}$$
$$\text{Total web stripped-off per stroke} = 0.55 + 0.125 = 0.675 \text{ in}$$
$$\text{Comb box speed} = (1188)/0.675 = 1760 \text{ strokes/min}$$

2. Assume cylinder speed to be 170 rev/min. The cylinder pulley of 20 in drives another small pulley of 15 in by means of an open belt. On the same shaft, another 10 in grooved pulley drives a 4-inch grooved pulley on the comb box pulley with a cross rope. How many oscillations per minute will the comb box make?

$$\text{Cycles per minute} = 170 \times \frac{20}{5} \times \frac{10}{4} 1700 \text{ strokes}$$

11.1.5.2 Exercises

1. Find the number of strokes of a doffer comb when the doffer makes 11 rpm. Total sweep is 1⅛ in. At the downward-most position, the comb moves away by 1/16 in from doffer and about 35% of the stroke is useful.
   [Ans: 2043 strokes/min]
2. A grooved pulley of 19 in on the cylinder drives a 5½ in grooved pulley. Another compound pulley of 9½ in drives a 3½ in grooved pulley on the doffer comb shaft. If the speed of the cylinder is 180 rpm, calculate the number of oscillations of doffer comb.
   [Ans: 1687 strokes/min]

11.1.6 CARD–DRAFT CALCULATIONS (WORKED EXAMPLES)

1. If it is desired to make a 60 grs/yd sliver from 14 oz/yd lap, what is the actual draft needed? If during processing the card extracts 3.5% waste, what is the mechanical draft?

$$\text{Hank of the sliver delivered} = (8.33)/60 = 0.139$$
$$\text{Hank of the lap fed to card} = 0.019/14 = 0.00136$$

Note: When the weight is expressed in grains per yard, the constant for finding out English hank (Ne) is 8.33; similarly, when the weight is expressed in ounces (oz) per yard, the constant of 0.019 is used for finding the hank. This can be verified in this example. As 16 oz make 1 lb, and this is equivalent to 7000 grains, then

$$\text{Lap weight of 14 oz/yd means } (14 \times 7000) \div 16 = 6125 \text{ grains/yd}$$
$$\text{Using the 8.33 constant, we have Hank of Lap} = 8.33/6125 = 0.00136$$

$$\text{Draft} = \frac{\text{Hank Delivered}}{\text{Hank Fed}} = \frac{0.139}{0.00136} = 102.21$$

This is the draft put in the carding machine to transform the lap into a sliver. However, as this thinning-out process also involves extraction of 3.5% waste, the resultant draft is referred to as *actual draft*. There exists a simple relation between actual draft and the one due to mechanical gear ratios of delivery speed to feeding speed (called mechanical draft).

$$\text{Mechanical draft} = \text{Actual draft} \times \frac{(100 - \text{waste\%})}{100} = 102.21 \times \frac{(100 - 3.5)}{100}$$

Therefore, mechanical draft = $(102.21 \times 96.5)/100 = 98.63$

Note: It can be easily remembered that **Actual Draft > Mechanical draft.**

2. The lap fed to a card has a mass of 600 g/m. The linear density of the card sliver is 5 ktex. What is the draft employed in the card?

Sliver of 5 ktex means that 1 kilometer of sliver weighs 5 kg

$$= 5000 \text{ g}/1000 \text{ m or} = 5 \text{ g/m}$$

$$\text{Draft} = \frac{\text{Weight fed}}{\text{Weight delivered}} = \frac{600}{5} = 120$$

Note: This is *actual draft* as realized during the thinning out of the material.

11.1.7 EXERCISES

1. If a card is fed with lap of 14 oz/yd and if the weight of the sliver delivered is 55 grs/yd, find the actual draft.
   [Ans: 111.36]
2. The actual draft in a card producing 50 grs/yd of sliver with 5% waste is 110. Calculate the mechanical draft.
   [Ans: 104.5]
3. The lap fed to the card has a mass of 500 g/m. The linear density of card sliver is 5.5 ktex. What is the draft employed?
   [Ans: 90.91]
4. The surface speed of a feed roller is 35 cm/min and that of a coiler calender roller is 41.3 m/min. What is the mechanical draft?
   [Ans: 118]

11.1.8 CARD PRODUCTION CALCULATIONS (EXAMPLES)

1. The production of a conventional card is 12 lb/h. There are 20 cards running to process similar material. The waste taken out at each card amounts to 5%, whereas the hank of the lap fed to the card is 0.0012. Calculate at what speed the 9 in lap rollers of scutcher should run with a working efficiency of 90% so as to meet the carding requirements.

Production for all the cards in the department = $12 \times 20 = 240$ lb/h

With 5% waste extracted at card, the back-stuff (feed) required will be
= $(240 \times 100)/95 = 252.63$ lb/h

0.0012 hank of blow room lap is equal to $1/(0.0012 \times 840) =$ approx. 0.991 lb/yd (**Note:** From basic definition of count: 840 yd in one pound)

This means that to produce 252.63 lb/h, the scutcher must produce 252.63/0.991 yd of lap length (i.e., 254.92 yd) in 1 hour or 4.248 yd/min. (11.4)

$$\text{Surface Speed of Lap Roller} = \frac{\pi \times D \times N}{36} \times \frac{90}{100} \qquad (11.5)$$

Where D = diameter of lap roller in inches and N = its rpm

We can equate expression (11.4) with (11.5). Thus,

$$4.248 = \frac{22 \times 9 \times N \times 90}{7 \times 36 \times 100} \text{ Hence, } N = \frac{4.248 \times 7 \times 36 \times 100}{22 \times 9 \times 90}$$

Therefore, N = 6 rpm, i.e., the lap roller of scutcher should run at 6 rpm so as to meet the required production for feeding 20 cards.

2. If the hank of the sliver at card is 0.12 and the actual draft in the machine is 104, calculate the weight of a lap measuring 40 yd.

$$\text{Actual Draft} = (\text{Hank Delivered})/\text{Hank Fed}$$

$$104 = 0.12/\text{Hank Fed}$$

Hence, the hank fed = 0.12/104 = 0.00115 = 1.035 lb/yd

The weight of the lap with 40 yd length will be (1.035 × 40) = 41.40 lb

3. A lap of 14 oz/yd is fed to a card with mechanical draft of 100. The total waste extracted at card is 5%. Calculate the hank of sliver and also the weight of 6 yd wrapping.

Actual or resultant draft considering the waste extracted will be

$$(100 \times 100)/(100 - 5) = 105.26$$

$$\text{Actual draft} = \frac{\text{Weight/yd of the Lap Fed}}{\text{Weight/yd of Sliver Delivered}} \text{ Wt/yd}$$

Hence, 105.26 = 14/(sliver weight delivered

Sliver produced = 14/105.26 = 0.133 oz/yd or 0.133 × 453.5 = 60.3 grs/yd

Note: 453.5 grains = 1 ounce (oz). Hence, weight/6 yd = 361.8 grs, which is equal to weight of wrapping.

Using a constant of 8.33, the hank of the sliver will be

$$= 8.33/60.3 = 0.138 \text{ hank}$$

4. If the feed roller in carding makes 3 rev/min and its diameter is 6 cm, what is the feed rate in terms of length and weight? The lap has a mass of 500 g/m and the draft between feed roller and lap roller is 1.04.

$$\text{Length fed by feed roller} = \pi \times 6 \times 3 = 56.55 \text{ cm or } 0.5655 \text{ m/min}$$

Depending upon the draft employed between the feed roller and lap roller, the lap weight fed at feed roller will have slightly less mass per unit length.

$$\text{Hence the mass fed at the feed roller} = 500/1.04 = 480.7 \text{ g/m}$$

Thus, the mass fed per unit time = 480.7 × 0.5655 = 271.87 g

5. Calculate the production of a card per shift of 8 hr if the draft between the doffer and the lap roller is 90. Further, the draft between calender roller and doffer is 1.02 and that between coiler calender roller is 1.05. Also, the hank of the lap fed is 0.00146, whereas the waste removed during the process is 4%. The doffer of 27 in diameter runs at 12 rpm with 90% efficiency.

$$\text{Total Draft} = 90 \times 1.02 \times 1.05 = 96.39$$

This is total mechanical draft. The actual total draft will be

$$(100 \times 96.39)/(100-4) = 100.41$$

Hence, the hank of the sliver as delivered by the coiler calender roller will be

$$= 0.00146 \times 100.41 = 0.146$$

Also, as the mechanical draft up to doffer is 90, the corresponding actual draft will be

$$(90 \times 100)/(100-4) = 93.75$$

Hence, the hank of material at doffer will be $0.00146 \times 93.75 = 0.1365$ (11.6)

The example can be solved in two ways: a) calculating the production from doffer speed with the hank of material as delivered at doffer, or (b) finding the final hank of the sliver as delivered by coiler calender roller and appropriately taking its delivery rate.

a. Thus, with the hank of lap fed as 0.00146, the hank of material at doffer as per (11.6),

$$\text{Production from doffer} = \frac{\pi \times 27 \times 12 \times 60 \times 8 \times 0.9}{36 \times 840 \times 0.1365} = 106 \text{ lb/shift} \qquad (11.7)$$

b. The production from the coiler calender roller can now be calculated. The surface speed of the coiler calender roller will be that of the doffer multiplied by the draft between coiler calender roller and doffer. Thus,

$$\text{Surface Speed of Coiler C.R.} = [(\pi \times 27 \times 12) \times 1.02 \times 1.05]/36$$
$$= 30.28 \text{ yd/min}$$

$$\text{Production of Card from Coiler C.R.} = \frac{30.28 \times 60 \times 8 \times 0.9}{840 \times 0.1465 *} = 106 \text{ lb/shift} \quad (11.8)$$

It can be seen that the values from (11.7 & 11.8) that are the same.

6. A conventional revolving flat card produces 140 lb per unit time with sliver of 70 grains/yd. The production change wheel has 26 teeth. What should be the new value of this change wheel, if the hank is made finer as 64 grains/yd so that the same production rate is maintained?

Present hank × Present change wheel = Required hank × Required change wheel $70 \times 26 = 64 \times$ Required change wheel

Thus, the new required change wheel would be $= (70 \times 26)/64 = 28.43$ or 28 teeth

11.1.9   EXERCISES

1. If the draft constant in the card is 1624, and the draft in the machine is 98, find the necessary draft change wheel. Further, if the draft change wheel is changed over to 24 teeth, what is the new draft value?

[Ans: 17^T approx.; 67.66 draft]

---

* Hank of sliver at coiler calender roller as per (11.6)

2. What should be the weight of lap suitable for a card, if the waste extracted is 5%? The mechanical draft employed in the machine is 120 and the required sliver weight is 36 grains/yd.

   [Ans: 10.39 oz/yd]

3. At what speed should the doffer be run in order to produce 400 lb of sliver in 54 hours? The weight per unit length of sliver is 34.6 grains/yd and the doffer has 24.75 in diameter. The draft between coiler calender roller and doffer is 1.1

   [Ans: 10.5 rpm]

4. The mechanical draft in a card is 102 and the sliver required to be produced is 54 grains/yd with 5% waste extracted. What is the hank of the lap that should be used?

   [Ans: 0.00143]

5. Find the production of 120 cards in 48 hours, if the doffer of 27 inches diameter makes 13 rev/min. The weight of the lap is 12 oz/yd and the mechanical draft up to doffer is 95. The waste extracted at the card is 5%, whereas the machine stoppage amount to 7% of the time.

   [Ans: 615.38 lb/card; 73,846.3 lb for 120 cards]

6. The surface speed of a coiler calender roller was found to be 39.1 m/min. If the linear density of the sliver is 4 ktex, what is the production per hour at 80% efficiency?

   [Ans: 7.5 kg/hr]

7. A card has a production of 120 lb per shift with a 28 teeth production change wheel. If the production rate required is 140 lb per shift, what is the new change wheel when the weight of the sliver per yard remains the same?

   Note: Present Production Rate × Required Change Wheel = Required Production Rate × Present Change Wheel

   [Ans: 32.66or approx. 33 teeth]

8. The production of a card is 120 lb per unit time; the sliver weight is 60 grains/yd and the production wheel has 20 teeth. What is the value of the production wheel to produce 110 lb in the same time and producing 50 grains/yd of sliver?

   Note: Present Production Rate × Required Sliver Weight × Present Change Wheel = Required Production Rate × Present Sliver Weight × Required Change Wheel

   [Ans: New Change Wheel = 18 teeth]

11.1.10   CALCULATIONS OF LENGTH AND COUNT OF FILLET

The conventional cards used to be clothed with flexible fillet wire clothing. The fillet is generally 2 in wide. The length of the fillet required to cover a width of 40 in and with cylinder of 50 in will be

$$= \frac{50 \times 40 \times 22}{2 \times 12 \times 7} = 261.90 \text{ ft}$$

It is possible to weigh a small length of the fillet and then find the total weight of this fillet. Similar types of calculations can be done for metallic wire with a rib width of 0.038 in. Thus the length of metallic fillet will be

$$= \frac{50 \times 40 \times 22}{0.038 \times 7 \times 36} = 4594.8 \text{ yd } [= 165{,}412.8 \text{ in}]$$

If the wire weighs 302 mg/in then the weight of metallic fillet for cylinder will be $[165{,}412.8 \times 302]/(1000 \times 1000) = 49.95$ kg

   a. The count of wire of card clothing in the American system is the number of wire points in a square foot of clothing.
   b. The count of card clothing in the English system is the number of wire points per square inch divided by 5.

11.1.10.1 Worked Examples

   1. Calculate the number of points per square foot and also the count of the fillet in English system, if crowns per 2 in are 8, nogg per inch are 23 and points per nogg are 3.

Crown per foot = $(8 \times 12)/2 = 48$. Further, as each crown has 2 points

Points per foot = $48 \times 2 = 96$

Nogg per foot = $23 \times 12 = 276$. Therefore, points per foot = $276 \times 3 = 828$

Hence,

$$\text{Points per square foot} = 828 \times 96 = 79{,}488$$

$$\text{Count English system} = \frac{828 \times 96}{12 \times 12 \times 5} = 110.4$$

Note: In the case of rib set fillet, the crowns extend (Figure 11.3) across the width of the fillet. Normally, there are four crowns per inch. Noggs run lengthwise on the fillet and vary from 10 to 28 per inch. One nogg consists of a group of 3 crowns. Cylinder fillet has 8 ribs, whereas the doffer fillet has 6 ribs. Nogg is the number of crowns set into the foundation in one repeat of the lengthwise pattern.

   The closeness of the points is thus expressed in noggs per inch and this forms the unit of its measure. In the case of twill set fillet, the crowns extend along the lengthwise strip and are expressed per inch. The noggs are counted across. In each nogg, there are 6 crowns. This type of fillet was used for flats.
   2. Cylinder fillet is 2 in wide, whereas doffer is 1½ in in width. In cylinder fillet, there are 8 ribs across the fillet (rib = crown) and crowns per inch = 4.

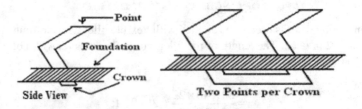

FIGURE 11.3   Flexible wires[l]: The specialty of flexible wires is the bent knee, which gives more piercing power. The semi-rigid foundation serves to give limited flexibility.

Calculate the total points per inch and also the count of fillet in the English system. Assume that there are 2 teeth per crown, 8 teeth per inch, 83 noggs per 4 in and 3 crowns per nogg.

$$\text{Noggs per inch} = 83/4 = 20.75$$

$$\text{Total points per inch} = \text{Crowns per inch} \times \text{Teeth/Crown} \times \text{Noggs/inch} \times \text{Crowns/Nogg}$$
$$= 4 \times 2 \times 20.75 \times 3 = 498$$

Therefore English Count = (498) ÷ 5 = 99.6 (or approx. 100^s)

11.1.10.2   Exercise

Find the count of a rib set fillet in which crowns per 2 inch are 8, noggs per inch are 11 and points per nogg are 6

[Ans: American system = 76,032; English = 105.6]

11.1.11   WASTE PERCENTAGE CALCULATIONS

Usually a full lap of known weight is fed to the card, after the whole lap is finished, the droppings are collected. Also, if possible, the weight of the entire sliver delivered is taken. This gives a fairly good idea of how much the card extracts the waste over a long time.

11.1.11.1   Worked Examples

1. A lap of 16 kg was fed and consumed at the card, and the waste collected at different regions was as follows: Licker-in Droppings = 300 g, Flat Strip = 360 g; Cylinder and Doffer fly = 30 g and Scavenger Roller = 10 g. The sliver collected was 15.00 kg. Calculate the individual and the total waste percentage and invisible loss.

$$\text{Licker-in droppings} = (0.300 \times 100)/16 = 1.87\%$$

$$\text{Flat strip} = (0.360 \times 100)/16 = 2.25\%$$

$$\text{Cylinder–doffer fly} = (0.030 \times 100)/16 = 0.18\%$$

$$\text{Scavenger roller} = (0.010 \times 100)/16 = 0.06\%$$

$$\text{Total } (0.700) \text{ Waste} = 4.36$$

Now,

Total waste collected = 700 g; therefore the sliver collected should have been 16.0 − 0.700 = 15.3 kg. However, the sliver collected was 15.0 kg. Hence, the invisible loss through the cages = 15.3 − 15.0 = 0.3 kg

When this is expressed as a percentage = (0.3 × 100)/16 = 1.87% (invisible loss)

2. One yard of lap weighs 1 pound. A sliver made of this lap had the weight of 55 grains/yd. The card extracts 5% waste. Calculate the draft.

   Note: The reduction from lap to sliver (1 lb/yd to 55 grains/yd) in the above case includes the waste% extracted and this reduction represents the actual draft.

$$\text{Actual draft} = 7000/55 = 127.27 \text{ [1 lb = 7000 grains]}$$

Now, taking waste extracted as 5%, we can calculate mechanical draft.

$$\text{Mechanical draft} = 127 \times 0.95 = 120.9$$

11.1.11.2 Exercises

1. A lap of 15 kg was fed to a card when the licker-in droppings were 2% and flat strip was 1.9%. If the total waste extracted was 620 g, find the invisible loss. Consider the cylinder and doffer fly and other waste negligible.
   [Ans: 0.23%]
2. A full lap of 18 kg was fed to a card when the total waste extracted was found to be 5%. If the invisible loss, out of the total waste, was 0.4% and the flat strip was 2.5%, find the weight of net card sliver produced and also the weight of licker-in droppings.
   [Ans: 17.1 kg & 378 g]

11.2   DRAW FRAME CALCULATIONS FROM GEARING

The gearing diagram is shown in Figure 11.4.

11.2.1   Speed Calculations

1. Front roller (FR)

$$= \frac{940 \times 4.25 \times 11}{1 \times 12 \times 11} = 332.9 \text{ rpm}$$

$$\text{Surface speed} = \frac{332.9 \times \pi \times 1\frac{3}{8}}{12} = 119.85 \text{ ft/min}$$

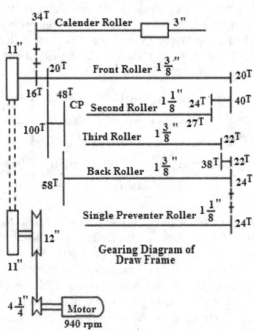

FIGURE 11.4   Draw frame gearing[1]: The gearing is perhaps the simplest of all the spinning machines. There are, however, very important wheels like change pinion, back roller wheel or even crown wheel; all of them control the draft in draw frame.

2. Back roller (BR)

$$= 332.9 \times \frac{20}{100} \times \frac{48}{58} = 55.1 \text{ rpm}$$

$$\text{BR surface speed} = \frac{55.1 \times \pi \times 1\frac{3}{8}}{12} = 19.84 \text{ ft/min}$$

3. Second roller (SR)

$$= 332.9 \times \frac{20 \times 24}{40 \times 27} = 147.95 \text{ rpm}$$

SR surface speed

$$= (147.95 \times \pi \times 1\frac{1}{8}) / 12 = 43.58 \text{ ft/min}$$

4. Third roller (TR)

$$= 55.1 \times \frac{24 \times 38}{22 \times 22} = 103.82 \text{ rpm}$$

$$\text{TR surface speed} = (103.82 \times \pi \times 1\frac{3}{8})/12 = 37.37 \text{ ft/min}$$

5. Calender roller (CR)

$$= 332.9 \times \frac{16}{34} = 156.65 \text{ rpm}$$

$$\text{CR surface speed} = (156.65 \times \pi \times 3)/12 = 123.0 \text{ ft/min}$$

11.2.2 Draft Calculations

1. Draft between FR & SR= 119.85/43.58 = 2.75
2. Draft between SR & TR = 43.58/37.37 = 1.17
3. Draft between TR & BR = 37.37/19.84 = 1.88 (Break draft)
4. Draft between CR & FR = 123.0/119.85 = 1.02 (Tension draft)
5. Total Draft between CR & BR = 123.0/19.84 = 6.20
6. Draft constant = Total draft × Change pinion (CP)

$$= 6.20 \times 48 = 297.6$$

11.2.3 Production Calculations (from Calender Roller)

It may be assumed here that the hank of the drawn sliver is 0.15 and the machine efficiency is 75%.

$$\text{Production per delivery} = \frac{123 \times 60 \times 8}{3 \times 840} \times \frac{0.75}{0.15 \times 2.205} = 53.12 \text{ kg/shift}$$

Production for drawing frame with 4 deliveries = 53.12 × 4 = 212.48 kg/shift

11.2.4 WORKED EXAMPLES

1. Total draft between front roller and back roller of drawing frame is 5.88. Calculate the intermediate drafts between each pair of rollers if the draft between front and the third roller is 4.65 and that between second and third roller is 45% greater than the draft between third and fourth roller.

Draft between I & III zone = 4.65

Hence the draft between III & IV zone = (Total draft) ÷ 4.65 = 5.88 ÷ 4.65 = 1.26

Draft between II & III zone will be = Draft in III & IV zone × 1.45 (45% greater)

$$= 1.26 \times 1.45 = 1.83$$

Hence the draft in I & II zone will be = (Draft in I & III zone)/(Draft in II & III zone)

$$= 4.65/1.83 = 2.54$$

2. A draw frame with 56 teeth draft change pinion delivers a sliver of 48 grains/yd from six slivers, each of 45 grains/yd. What should be the change pinion required, if the sliver produced needs to be of 52 grains/yd?

Presently, the actual draft = (45×6) / 48 = 5.625   (11.9)

Note: Draft = (Sliver Weight Fed × No. of Doublings) ÷ (Sliver Weight Delivered)

Required draft = (45×6) ÷ 52 = 5.19   (11.10)

Draft constant = Present draft × Present change pinion (CP) = 5.625×56 = 315

Required CP = Draft constant/Required draft

$$= 315/5.19 = 60.69 \text{ Approximately} = 61 \text{ teeth}$$

3. Calculate the production per shift of 8 hours of draw frame running at 85% efficiency. The front roller of 1¼ in diameter runs at 300 rpm. The weight of the sliver fed is 60 grains/yd and the draft employed with six doublings is 6.2. The machine has six deliveries.

Weight per yard of sliver delivered = (60 × 6) ÷ 6.2 = 58.06 grains/yd

$$\text{Production/shift} = \frac{22 \times 5 \times 300 \times 60 \times 8}{7 \times 4 \times 36} \times \frac{58.06}{7000} \times 0.85 \times 6$$

$$= 664.72 \text{ lb or } 301.46 \text{ kg}$$

4. Calculate the efficiency of draw frame where the production for two machines with 4 deliveries each is 4124 lb for a time period of 40 hours. The calender roller of 2 in diameter runs at 200 rpm and delivers a sliver of 50 grains/yd.

Actual production for 8 deliveries = 4124 lb

Theoretical production from the calender roller producing the sliver will be

$$\text{Theoretical production} = \frac{22 \times 2 \times 200 \times 60 \times 40 \times 8}{7 \times 36} \times \frac{50}{7000}$$

Theoretical Production = 4789 lb

Actual production

$$\text{Efficiency} = \frac{\text{Actual Production}}{\text{Theoretical production}} \times 100 = \frac{4124 \times 100}{4789} = 86.1\%$$

5. If the production of a drawing frame with six deliveries is 920 lb in a shift of 8 hours and sliver delivered is 60 grains/yd, calculate the speed of calender roller of 1¼ in diameter. The machine efficiency is 85%.

$$\text{Prod/shift} = \frac{\pi \times \text{FR dia.} \times \text{FR rpm} \times 60 \times 8 \times \text{grs/yd} \times \text{No. of doublings} \times \text{Efficiency}}{\text{in/yd} \times \text{grains/lb}}$$

$$920\,\text{lb} = \frac{22 \times 5 \times \text{FR rpm} \times 8 \times 60}{7 \times 4 \times 36} \times \frac{60}{7000} \times 6 \times 0.85$$

Hence,

$$\text{FR speed} = 401.8\ \text{rpm}$$

6. Calculate the number of deliveries of a draw frame required to produce 30,000 lb of sliver per week of 40 hours, when sliver of 50 grains/yd is delivered. The calender roller has a 2 in diameter and runs at 260 rpm. The overall machine efficiency is 78%.

$$30,000 = \frac{22 \times 2 \times \pi}{7 \times 12 \times 3} \times \frac{260 \times 60 \times 40 \times 50}{7000} \times 0.78 \times \text{No. of deliveries}$$

$$30,000 = 2471.88 \times \text{No. of deliveries}$$

Rearranging the expression for the required number of deliveries, we get Number of draw frame deliveries required to produce 30,000 lb will be

$$= 30,000.0 / 2471.88 = 12.13 = \text{approximately 12 deliveries}$$

Thus, 3 (conventional) draw frames, each with 4 deliveries, will be required.

7. A conventional 4/4 draw frame has following particulars:
   a. Front roller diameter = 1⅛ in (28.6 mm)
   b. Front roller speed = 220 rpm
   c. Tension draft between coiler CR & FR = 1.05
   d. Hank of sliver delivered = 0.16 Ne

Calculate the production of the draw frame in kg per 7.5 hours with 85% efficiency.

Number of hanks of 840 yd delivered by the front roller in a shift of 7½ hours

$$= \frac{9 \times \pi \times 200 \times 60 \times 7.5 \times 0.85}{8 \times 12 \times 3 \times 840} = 8.94\ \text{Hanks}$$

$$\text{Production} = \frac{\text{Hanks}}{\text{Hank of Sliver}} = \frac{8.94}{0.16} = 55.87\ \text{lb or 25.33 kg/shift}$$

As there is draft of 1.05 between calender roller and front roller; the delivery rate will have to take this draft into account. This is because the hank of the sliver as 0.16 is actually at the calender roller.

$$\text{Production} = 25.31 \times 1.05 = 26.57\ \text{kg}$$

8. In a mill, there are two draw frame passages employed. At every passage, there are six slivers doubled. By oversight, the draft in the third passage was changed from 6.0 to 5.4. What will be the percentage change in the observed production rate?

Production is inversely proportionate to draft.

Thus, when the draft is reduced, the production will increase.

This increase will be by 6.0/5.4 times. Therefore, if P is the original production, the new production will be,

$$(6/5.4) \times P = 1.111 \, P$$

The increase in the production = $1.111 \, P - P = 0.111 \, P$

By percentage, this will be equal to $[(0.111 \, P)/P] \times 100 = 11.11\%$

9. If the surface speed of the front roller of draw frame is 200 m/min and the draft between coiler calender roller and front roller is 1.05, calculate the production per shift of 8 hours with 85% efficiency. The sliver produced is 5 ktex.

The surface speed of coiler CR = $200 \times 1.05 = 210$ m/min

$$\text{Production} = \frac{210 \times 60 \times 8 \times 0.85 \times 5}{1000} = 428.4 \text{ kg/shift/delivery}$$

With two deliveries per machine, production = $428.4 \times 2 = 856.8$ kg

11.2.5 CALCULATIONS FOR HIGH-SPEED DRAW FRAMES

It is now known that for calculating production, basically three things are required:

1. Delivery rate
2. Hank of sliver
3. Machine efficiency

Whenever necessary, the tension drafts must also be taken in to account. For any high-speed draw frame, this draft never exceeds a value of 1.05 (sometimes even less) and for the sake of calculations, the value may lie between 1.02 and 1.03.

1. The front roller of Lakshmi Rieter draw frame runs at 1960 rpm and its diameter is 38 mm. The tension draft between coiler calender roller and front roller is 1.03. Calculate the production per machine for 8 hours, with efficiency as 80% and hank of the sliver produced as 0.13

$$\text{English hank delivered} = \frac{1960 \times 38 \times \pi \times 60 \times 8}{10 \times 2.54 \times 36 \times 840} = 146.24 \text{ hanks/shift}$$

At 80% efficiency, the hanks delivered will be = $146.24 \times 0.8 = 116.99$

$$\text{Production per delivery per shift} = \frac{116.99}{0.13 \times 2.205} = 408.14 \text{ kg}$$

For two delivery machines, the production = $408.14 \times 2 = 816.28$ kg per shift

It may be noted that this production is calculated from front roller delivery. Now taking into account the tension draft between coiler calender roller and front roller, the actual hanks delivered by the machine will be proportional tension draft. This will correspondingly increase the net length delivered by the machine. The production will also increase by the same proportion. Thus, the production at the coiler calender roller = 816.28 × 1.03 = 840.76 kg

2. A Whitin draw frame has front roller of 2 in which runs at 1400 rpm. The tension draft between coiler calender roller and front roller is 1.02. Calculate the production at 80% efficiency when the hank delivered is 0.15.

$$\text{Number of hanks delivered} = \frac{1400 \times 2 \times \pi \times 60 \times 8}{36 \times 840} \times 0.8 \times 1.02 = 113.94 \text{ hanks/shift}$$

$$\text{Production per delivery} = \frac{113.94}{0.15 \times 2.205} = 344.51 \text{ kg/shift}$$

Thus, the machine with 4 deliveries will give 345.37 × 4 = 1381.48 kg/shift

3. In a Saco-Lowell draw frame, a draft change wheel has 55 teeth. The sliver fed and delivered are 44 grains and 48 grains/yd, respectively. Eight slivers are fed to the machine. What is the change wheel required to get the sliver of 53 grains/yd?

$$\text{The draft in the draw frame} = (44 \times 8) \div 48 = 7.33$$
$$\text{The draft constant will thus} = \text{draft} \times \text{CP} = 7.33 \times 55 = 403.15$$

Now to get 53 grains/yd, the draft required will be

$$= (44 \times 8)/53 = 6.64$$

Therefore the change pinion required will be = Draft constant/New draft
$$= 403.15/6.64 = 60.71$$
$$= \text{Approximately 61 teeth}$$

11.2.6 EXERCISES

1. In a draw frame with metallic fluted top rollers, the 1⅛ in front roller runs at 320 rpm. The weight of the sliver fed is 60 grains/yd and six slivers are doubled. The draft in the machine is 5.8 whereas the tension draft of 1.02 is put between coiler calender roller and front roller. Calculate the production per shift of 8 hours per delivery, assuming crimp of 20% and efficiency of 80%. [20% crimp means 100 yd will be delivered as 120 yd]
   [Ans: 78.97 kg/delivery]
2. Surface speed of back roller of a conventional draw frame is 20 ft/min. Six slivers, each of 50 grains/yd are fed. What will be the production, if the hank delivered is the same?
   [Ans: 62.17 kg/delivery/shift]
3. A conventional draw frame has front roller of 1⅜ in diameter and it runs at 420 rpm. The hank delivered is 0.14. Calculate the production per machine with 4 deliveries and at 70% efficiency.
   [Ans: 65.30 kg/delivery]

4. Lakshmi Rieter draw frame working at 635 ft/min delivers 0.16 hank of sliver at 78%. Find the production per shift of 8 hours.

   [Ans: 267.41 kg/delivery]

5. Whitin draw frame with two deliveries has a front roller of 2 in diameter and it runs at 1700 rpm. The hank delivered is 0.13. Calculate the production per shift of 8 hours at 75% efficiency.

   [Ans: 443.66 kg/delivery]

6. With a front roller speed of 415 rpm, calculate the production of a draw frame for a shift of 8 hours, when the front roller diameter is 1⅛ in and the hank delivered is 0.16. The machine works at 85% efficiency.

   [Ans: 65.99 kg]

7. A high-speed draw frame produces 500 kg per shift per delivery with 85% efficiency and 0.15 hank of sliver. If the coiler calender roller diameter is 2.75 in, find its rpm.

   [Ans: 1418.57 rpm]

8. The coiler calender roller in a draw frame runs at 1500 m/min. The machine has a draft constant of 340 and 40 teeth CP is being worked. The feeding cans have 4000 m of sliver coiled. Find the time to consume the feed cans.

   [Ans: 22.66 min]

REFERENCE

1. Elements of Cotton Spinning, Carding & Draw Frame – Dr. A.R. Khare, Sai Publication, 1999, Mumbai

Index

Printed in the United States
by Baker & Taylor Publisher Services